Analog Circuits and Signal Processing

Series Editors:

Mohammed Ismail, The Ohio State University
Mohamad Sawan, École Polytechnique de Montréal

For further volumes:
http://www.springer.com/series/7381

Viranjay M. Srivastava • Ghanshyam Singh

MOSFET Technologies for Double-Pole Four-Throw Radio-Frequency Switch

Viranjay M. Srivastava
Assistant Professor
Department of Electronics
 and Communication Engineering
Jaypee University of Information
 Technology
Solan, Himachal Pradesh
India

Ghanshyam Singh
Professor
Department of Electronics
 and Communication Engineering
Jaypee University of Information
 Technology
Solan, Himachal Pradesh
India

ISBN 978-3-319-01164-6   ISBN 978-3-319-01165-3 (eBook)
DOI 10.1007/978-3-319-01165-3
Springer Cham Heidelberg New York Dordrecht London

Library of Congress Control Number: 2013946571

© Springer International Publishing Switzerland 2014

This work is subject to copyright. All rights are reserved by the Publisher, whether the whole or part of the material is concerned, specifically the rights of translation, reprinting, reuse of illustrations, recitation, broadcasting, reproduction on microfilms or in any other physical way, and transmission or information storage and retrieval, electronic adaptation, computer software, or by similar or dissimilar methodology now known or hereafter developed. Exempted from this legal reservation are brief excerpts in connection with reviews or scholarly analysis or material supplied specifically for the purpose of being entered and executed on a computer system, for exclusive use by the purchaser of the work. Duplication of this publication or parts thereof is permitted only under the provisions of the Copyright Law of the Publisher's location, in its current version, and permission for use must always be obtained from Springer. Permissions for use may be obtained through RightsLink at the Copyright Clearance Center. Violations are liable to prosecution under the respective Copyright Law.

The use of general descriptive names, registered names, trademarks, service marks, etc. in this publication does not imply, even in the absence of a specific statement, that such names are exempt from the relevant protective laws and regulations and therefore free for general use.

While the advice and information in this book are believed to be true and accurate at the date of publication, neither the authors nor the editors nor the publisher can accept any legal responsibility for any errors or omissions that may be made. The publisher makes no warranty, express or implied, with respect to the material contained herein.

Printed on acid-free paper

Springer is part of Springer Science+Business Media (www.springer.com)

Contents

1 Introduction 1
   1.1 Transceiver Systems 2
   1.2 Radio-Frequency Switches 5
   1.3 Radio-Frequency MOSFETs 7
   1.4 Issues of Radio-Frequency MOSFET Modeling 9
   1.5 Double-Gate MOSFET 13
   1.6 Cylindrical Surrounding Double-Gate MOSFET 15
   1.7 Hafnium Dioxide-Based MOSFET 16
   1.8 Image Acquisition of the MOSFETs 16
   1.9 Conclusion 16
   References 17

2 Design of Double-Pole Four-Throw RF Switch 23
   2.1 Introduction 23
   2.2 Comparison of Various Switches 23
      2.2.1 PIN Diode Switch 24
      2.2.2 GaAs FET Switch 24
      2.2.3 MESFET Switch 24
      2.2.4 MOSFET Switch 25
      2.2.5 MEMS Switch 25
   2.3 RF Transceiver Systems 26
   2.4 RF Transceiver Switch 28
   2.5 Design of CMOS Inverter for RF Switch 29
   2.6 Configuration of Switches 31
      2.6.1 Single-Pole Single-Throw Switch 31
      2.6.2 Single-Pole Double-Throw Switch 32
      2.6.3 Double-Pole Double-Throw Switch 33
      2.6.4 Double-Pole Four-Throw Switch 33
   2.7 Design of DP4T RF Switch Based on Single-Gate MOSFET 34
   2.8 Operational Characteristics of DP4T CMOS Switch 36

		2.9	RF Switch Performance Parameters	37
			2.9.1 Insertion Loss	38
			2.9.2 Return Loss	38
			2.9.3 Isolation	38
			2.9.4 RF Power Handling	38
			2.9.5 Linearity	39
			2.9.6 Transition Time	39
			2.9.7 Switching Speed	39
		2.10	Topologies for DP4T Switches	39
		2.11	Conclusions	40
		References		41
3	**Design of Double-Gate MOSFET**			45
		3.1	Introduction	45
		3.2	Design Process of Double-Gate MOSFET	48
		3.3	Effects of Double-Gate MOSFET on the Leakage Currents	50
			3.3.1 Subthreshold Leakage	51
			3.3.2 Gate Leakage	51
			3.3.3 Band to Band Tunneling of Electrons	51
		3.4	Performance Improvement of DG MOSFET over SG MOSFET	53
		3.5	Resistive and Capacitive Model of DG MOSFET and SG MOSFET	56
		3.6	Characteristics of the DG MOSFET with Aspect Ratios	63
		3.7	Design of DG MOSFET with Several Gate-Fingers	66
		3.8	Model of Series and Parallel Combination for Double-Gate MOSFET	73
		3.9	Conclusions	75
		References		76
4	**Double-Pole Four-Throw RF Switch Based on Double-Gate MOSFET**			85
		4.1	Introduction	85
		4.2	Basics of Radio System Design	85
			4.2.1 Path Loss	85
			4.2.2 Gain Cascade	86
			4.2.3 1 dB Compression Point	86
			4.2.4 Third-Order Intercept Point	87
			4.2.5 Thermal Noise	87
			4.2.6 Noise Figure	88
			4.2.7 Phase Noise	88
		4.3	Design of DP4T DG RF CMOS Switch	88
		4.4	Characteristics of DP4T DG RF CMOS Switch	90

4.5	Effective ON-State Resistance of DP4T DG RF CMOS Switch		94
	4.5.1	Parallel Combination of Resistance in a Device	96
	4.5.2	Choosing Transistor with Large Mobility	96
	4.5.3	Keeping $V_{gs}-V_{th}$ Large	96
	4.5.4	Aspect Ratio of a Transistor	96
4.6	Attenuation of DP4T CMOS Switch		97
	4.6.1	Causes of Attenuation	100
	4.6.2	Counteracting Attenuation	100
4.7	OFF-Isolation		101
4.8	Resistive and Capacitive Model of DP4T DG RF CMOS Switch		101
4.9	Switching Speed of DP4T DG RF CMOS Switch		103
4.10	S-Parameters of DP4T DG RF CMOS Switch		103
4.11	Conclusions		106
References			107

5 Cylindrical Surrounding Double-Gate RF MOSFET ... 111

5.1	Introduction		111
5.2	Analysis of CSDG RF MOSFET		114
5.3	Fabrication Process for CSDG RF MOSFET		117
5.4	Characteristics of CSDG MOSFET		118
5.5	Resistive and Capacitive Model of the CSDG MOSFET		122
5.6	Explicit Model of CSDG MOSFET		129
5.7	Gate Leakage Current, Noise Model, and Short Channel Effects for CSDG MOSFET		131
5.8	Cross talk in CSDG MOSFET Model		132
5.9	Advantages of the CSDG MOSFET Model		135
5.10	Conclusions		137
References			138

6 Hafnium Dioxide-Based Double-Pole Four-Throw Double-Gate RF CMOS Switch ... 143

6.1	Introduction		143
6.2	MOSFET Model with HfO_2		146
6.3	Fabrication Process of HfO_2-Based DG MOSFET		147
6.4	Parameters of HfO_2-Based MOSFET		149
	6.4.1	Oxide Capacitance per Unit Area	149
	6.4.2	Threshold Voltage	149
	6.4.3	Drain Currents	149
	6.4.4	Body Bias Effect	150
	6.4.5	Resistances	150
	6.4.6	Capacitances	150
	6.4.7	Figure of Merit	151
6.5	Switching Characteristics of HfO_2-Based MOSFET		151
	6.5.1	Fall Time	151
	6.5.2	Rise Time	152

		6.5.3	Maximum Signal Frequency	152
		6.5.4	Propagation Delay	152
		6.5.5	Power Dissipation	153
	6.6	DP4T Switch Design with HfO$_2$-Based DG MOSFET		153
	6.7	Characteristics of DP4T Switch with HfO$_2$-Based DG MOSFET		155
		6.7.1	Drain Current Analysis	156
		6.7.2	ON/OFF Ratio and Insertion Loss	156
		6.7.3	ON-Resistance (R_{ON}) and Attenuation	157
		6.7.4	Flat-Band Capacitance and Dynamic Power	159
		6.7.5	Debye Length Calculation and Mobility	159
		6.7.6	Potential Barrier	160
	6.8	Conclusions		160
	References			161
7	**Testing of MOSFETs Surfaces Using Image Acquisition**			165
	7.1	Introduction		165
	7.2	Proposed Model for the Image Acquisition of MOSFETs		166
		7.2.1	Preprocessing	167
		7.2.2	Image Sensor	167
		7.2.3	Discrete Fourier Transform	169
		7.2.4	Filter Function	169
		7.2.5	Inverse Discrete Fourier Transform	170
		7.2.6	Postprocessing	170
		7.2.7	Image Enhancement	170
	7.3	Image Analysis		171
	7.4	Conclusion		172
	References			173
8	**Conclusions and Future Scope**			177
	8.1	Conclusions		177
	8.2	Future Scope		179
	References			181

Appendix A	**List of Symbols**	183
Appendix B	**List of Definitions**	187
Appendix C	**Outcomes of the Book**	191

About the Authors ... 193

Index ... 195

List of Figures

Fig. 1.1	Simple RF transceiver architecture	3
Fig. 2.1	Radio-frequency design hexagon	26
Fig. 2.2	A radio front-end block diagram with (**a**) the integration of transceiver switch and matching networks, (**b**) simplified schematic of a transceiver switch, and (**c**) typical transistor based transceiver switch	27
Fig. 2.3	Schematic of the CMOS (**a**) internal structure and (**b**) inverter circuit	30
Fig. 2.4	Schematic of the (**a**) SPDT, (**b**) DPDT, and (**c**) DP4T	32
Fig. 2.5	DP4T CMOS transceivers switch with single-gate transistor	34
Fig. 2.6	Schematic of the (**a**) basic SG MOSFET and (**b**) DP4T SG RF CMOS switch	35
Fig. 2.7	Proposed DP4T switch with two transistors	35
Fig. 2.8	Proposed DP4T switch layout with two transistors	37
Fig. 3.1	Schematic of the basic n-type double-gate MOSFET	46
Fig. 3.2	Layout of (**a**) DG MOSFET and (**b**) SG MOSFET	54
Fig. 3.3	Output voltage with gate and control voltage of (**a**) DG MOSFET and (**b**) SG MOSFET	55
Fig. 3.4	Drain current characteristics of (**a**) DG MOSFET and (**b**) SG MOSFET	57
Fig. 3.5	Voltage gain of (**a**) DG MOSFET and (**b**) SG MOSFET	58
Fig. 3.6	Layout of (**a**) n-type DG MOSFET and (**b**) p-type DG MOSFET	59
Fig. 3.7	The Circuit Models of (**a**) DG MOSFET and (**b**) SG MOSFET operating as a switch at ON-state	60
Fig. 3.8	Effect of the aspect ratio (when it is 2000) on the characteristics of DG MOSFET (**a**) drain current with gate to source voltage and (**b**) threshold voltage with the length (nm) of the channel	64

Fig. 3.9	Effect of the aspect ratio (when it is 500) on the characteristics of DG MOSFET (**a**) drain current with gate to source voltage and (**b**) threshold voltage with the length (nm) of the channel...	65
Fig. 3.10	Characteristics of capacitances with drain to source voltage for n-type MOSFET with the aspect ratio 2,000....................	66
Fig. 3.11	Layout of n-type DG MOSFET for (**a**) $NF = 1$ and (**b**) $NF = 10$..	67
Fig. 3.12	Voltage characteristice of n-type DG MOSFET for (**a**) $NF = 1$ and (**b**) $NF = 10$...................................	68
Fig. 3.13	Drain current characteristics of n-type DG MOSFET for (**a**) $NF = 1$ and (**b**) $NF = 10$...................................	69
Fig. 3.14	Output voltage characteristics of n-type DG MOSFET for (**a**) $NF = 1$ and (**b**) $NF = 10$...................................	70
Fig. 3.15	Conversion of the series and parallel combination of n-MOSFET/p-MOSFET to DG MOSFET. Case 1. Series combination of n-MOSFET to DG MOSFET. Case 2. Series combination of p-MOSFET to DG MOSFET. Case 3. Parallel combination of n-MOSFET to DG MOSFET. Case 4. Parallel combination of p-MOSFET to DG MOSFET...	74
Fig. 4.1	1 dB compression point...	87
Fig. 4.2	Third-order intercept point...	87
Fig. 4.3	Proposed DP4T DG RF CMOS switch.............................	89
Fig. 4.4	Layout of the proposed DP4T DG RF CMOS switch.............	91
Fig. 4.5	Characteristics of the proposed DP4T DG RF CMOS transceiver switch such as (**a**) applied input voltages, (**b**) antenna voltage with input voltages, (**c**) drain current, and (**d**) antenna output at various frequencies.....................	92
Fig. 4.6	Equivalent capacitive model of the proposed DP4T DG RF CMOS switch..	95
Fig. 4.7	Attenuation at $V_{CTL} = 0.7$–1.2 V for (**a**) 0.8-µm technology and (**b**) 45-nm technology................................	98
Fig. 4.8	Attenuation at $V_{CTL} = -0.1$ V to 0.7 V for (**a**) 0.8-µm technology and (**b**) 45-nm technology................................	98
Fig. 4.9	Resistive and capacitive model of DP4T DG RF CMOS switch at ON-state...	102
Fig. 4.10	Equivalent capacitive circuit of the DP4T DG RF CMOS switch..	104
Fig. 5.1	Schematic of (**a**) basic DG MOSFET, (**b**) CSDG MOSFET, and (**c**) cross-section of CSDG MOSFET........................	116
Fig. 5.2	Model of CSDG MOSFET transistor with its components at ON-state..	122

List of Figures   xi

Fig. 5.3	Design of the CSDG MOSFET with SPICE (**a**) capacitive models operating as a switch at ON-state, (**b**) input signal applied to gates, (**c**) output signal at drain, (**d**) source current variation with frequency, and (**e**) drain current variation with frequency..	124
Fig. 5.4	Design of the CSDG MOSFET with ADS (**a**) capacitive models operating as a switch at ON-state, (**b**) input signal applied to both gates, and (**c**) output signal at drain...............	126
Fig. 5.5	Equivalent resistive and capacitive model of the CSDG MOSFET...	133
Fig. 5.6	(**a**) Substrate cross talk mechanism and (**b**) Reduction of cross talk with CSDG MOSFET model.........................	134
Fig. 6.1	Dielectric constant vs. bandgap for gate oxides....................	144
Fig. 6.2	Schematic of the basic n-type MOSFET (**a**) with HfO_2 and (**b**) HfO_2 film on Si-substrate..................................	147
Fig. 6.3	Schematic of n-type DG MOSFET with HfO_2.....................	148
Fig. 6.4	DP4T RF CMOS switch with HfO_2-layered double-gate MOSFET..	154
Fig. 6.5	ON/OFF ratio for the proposed DP4T RF CMOS switch.........	157
Fig. 6.6	Attenuation for the proposed DP4T RF CMOS switch with respect to the applied control voltage.........................	158
Fig. 6.7	Insertion loss for the proposed DP4T RF CMOS switch with the ON-state resistance...	158
Fig. 7.1	Flow chart of a device testing using Image Acquisition..........	168
Fig. 7.2	Various images which can be obtained from the image acquisition of DG MOSFET...	171
Fig. 7.3	Various images which can be obtained from the image acquisition of CSDG MOSFET......................................	172

List of Tables

Table 1.1	Electromagnetic radiation spectrum	2
Table 1.2	Electromagnetic radiation spectrum based on IEEE	3
Table 3.1	Comparison of the various circuit parameters of the DG and SG MOSFET for proposed model	61
Table 3.2	Comparison of the drain current for proposed DG MOSFET model with the existing model	62
Table 3.3	Comparison of the various circuit parameters of the DG MOSFET for $NF = 1$ and $NF = 10$	71
Table 3.4	Design for independent gate configuration (IGC) and tied gate configuration (TGC)	74
Table 3.5	An effective aspect ratio for different combination of transistors as shown in Fig. 3.15	75
Table 4.1	Simulation results for drain current and switching speed for several switches	94
Table 4.2	Comparison of the switching speed	94
Table 4.3	Performance parameters of the double-gate MOSFET transceiver switch	95
Table 4.4	DP4T DG RF CMOS switch attenuation for control voltage range 0.7 V–2.1 V	100
Table 4.5	DP4T DG RF CMOS switch attenuation for control voltage range –0.1 V to 0.7 V	100
Table 4.6	Simulated parameters of the DP4T DG RF CMOS switch	104
Table 4.7	Impedance, admittance, series equivalent, and parallel equivalent circuit parameters of the proposed switch	104
Table 4.8	S-parameters of a designed switch at various frequencies (Mag. = magnitude, Ang. = angle)	105
Table 4.9	Magnitude of S_{12} and S_{21} (both are equal) at various frequencies	105

Table 5.1	Comparison of the various circuit parameters of the CSDG MOSFET and existing CSSG MOSFET model	129
Table 5.2	Advantage of the proposed CSDG MOSFET model over several reported literatures for CSSG MOSFET	136
Table 6.1	Properties of hafnium dioxide	144
Table 6.2	Dielectric constant, bandgap, and conduction band offset on Si of the candidate gate dielectrics	148
Table 6.3	Comparison of parameters of HfO_2-based MOSFET with the SiO_2-based MOSFET	153
Table 6.4	Working functionality of DP4T RF CMOS switch with HfO_2 DG MOSFET	155

Abstract

With the development of modern silicon technology, more and more high-frequency circuits can be implemented in standard complementary metal-oxide-semiconductor (CMOS) processes. The feasibility of RF ICs in standard CMOS process is established, and the trend in putting all components of a system on a chip includes integration of the transceiver (T/R) antenna switch.

In this book, we have designed a double-gate (DG) MOSFET and double-pole four-throw (DP4T) RF switch to enhance its performance for the next generation wireless communication systems. Further we have combined the ideas of DG MOSFET and DP4T switch to design a novel DP4T DG RF CMOS switch. The designed DP4T DG RF CMOS switch can route four inputs to two outputs at a time or vice versa. So it is twice effective as compared to the previously existing SPDT switches.

In the DG MOSFET, the gates are only on the two sides of the substrate. Hence, to utilize all the sides of the substrate, we have widened the gate all around the device and designed like a cylinder. Therefore, we extend this work to the cylindrical surrounding double-gate (CSDG) MOSFET. It has less contact area with the board compared to the other MOSFETs. Due to the circular source and drain, the gate contact with the source and drain is on a long circular region, which avoids the gate misalignment. This work has been extended by replacing SiO_2 with HfO_2 as a high dielectric material to design DG MOSFET.

Finally, we have analyzed the image acquisition of DG MOSFET and CSDG MOSFETs for the purpose of RF switch. The proposed model emphasized on the basics of single image sensor for two-dimensional images of a three-dimensional device, so that we can obtain a satisfactory device parameter.

Chapter 1
Introduction

With the development of electric telegraph by William Cooke and Charles Wheatstone, the telecommunication technology has been commercialized in 1838 [1]. This technology was rapidly replaced by Samuel Morse, with the introduction of the *Morse code* in 1844, which reduced the communication into dots and dashes, and listening to the receiver [2]. The wireless technology came to existence in 1901 when Guglielmo Marconi successfully transmitted radio signals across the Atlantic Ocean. The possibility of replacing the telegraphs and telephone communications with wave transmission is an exciting future. However, the two-way wireless communication has been materialized in the military, although it remained limited to one-way radio and television broadcasting by large and expensive stations. The ordinary two-way phone conversations would still go over wires for many decades. The invention of the large-scale integration (LSI) transistor, the development of Shannon's information theory, and the conception of the cellular system all at Bell Laboratories paved the way for affordable mobile communications.

The end of the twentieth century is remembered for the amazing growth of the telecommunication industry. The main cause for this event is the introduction of digital signal processing in the wireless communications, driven by the development of high-performance low-cost CMOS technologies for very-large-scale integration (VLSI). However, the radio-frequency (RF) analog front end remains the bottleneck for low-cost RF systems. The RF front-end design is pushed towards higher levels of integration and integration in low-cost CMOS technology, rendering significant space, cost, and power reductions. The cellular phones are no doubt the most popular wireless communication device currently in use. However, such a system can be divided into the user part (handset) and the infrastructure part (base stations). The user part consists of a transmitter and a receiver commonly known as transceiver system [3, 4].

The radio spectrum refers to the part of the electromagnetic spectrum corresponding to the radio frequencies (below 300 GHz). However, different parts of the radio spectrum are used for different radio transmission technologies and applications. The radio spectrum is typically government regulated in the developed countries and is sold or licensed to operators of private radio

Table 1.1 Electromagnetic radiation spectrum

Frequency	Abbreviation	Frequency range	Wavelength
Tremendously low frequency	TLF	Below 3 Hz	Above 10^5 km
Extremely low frequency	ELF	3–30 Hz	10^4–10^5 km
Super low frequency	SLF	30–300 Hz	10^3–10^4 km
Ultra low frequency	ULF	300–3,000 Hz	100–10^3 km
Very low frequency	VLF	3–30 kHz	10–100 km
Low frequency	LF	30–300 kHz	1–10 km
Medium frequency	MF	300 kHz–3 MHz	100 m–1 km
High frequency	HF	3–30 MHz	10–100 m
Very high frequency	VHF	30–300 MHz	1–10 m
Ultra high frequency	UHF	300 MHz–3 GHz	10 cm–1 m
Super high frequency	SHF	3–30 GHz	1–10 cm
Extremely high frequency	EHF	30–300 GHz	1 mm–1 cm
Tremendously high frequency/far infrared	THF/THz/FIR	300 GHz–3 THz	0.1–1 mm
Mid infrared	MIR	3–30 THz	10 μm–0.1 mm
Near infrared	NIR	30–300 THz	1–10 μm
Near ultraviolet	NUV	300 THz–3 PHz	0.1–1 μm
Extreme ultraviolet	EUV	3–30 PHz	10 nm–0.1 μm
Soft X-rays	SX	30–300 PHz	1–10 nm
Soft X-rays	SX	300 PHz–3 EHz	0.1–1 nm
Hard X-rays	HX	3–30 EHz	10 pm–0.1 nm
Gamma rays	Y	30–300 EHz	1–10 pm

transmission systems for the purpose of telecommunication or broadcast for television stations. To prevent interference and allow for efficient use of the radio spectrum, similar services are allocated in the bands. For example, the broadcasting, mobile radio, or navigation devices will be allocated in non-overlapping ranges of frequencies. However, each frequency range/band behaves differently and performs different functions and shared by civil, government, and military users of all nations according to International Telecommunications Union (ITU) radio regulations. For the communication purposes, the usable frequency range is in the range from 3 Hz to 300 GHz. In some cases, 100 THz is used for research purposes. These ranges of the frequency bands are given in Tables 1.1 and 1.2. The frequency band standards are also available in International Telecommunications Union radio regulations.

1.1 Transceiver Systems

With the development of wireless communication technology, the demand of high data rate wireless local area network (WLAN) systems is growing rapidly. The heterodyne receiver architecture is the most commonly used receiver architecture in the wireless communication systems. Due to the reduction of complexity and power

1.1 Transceiver Systems

Table 1.2 Electromagnetic radiation spectrum based on IEEE

Frequency	Abbreviation	Frequency range
High frequency	HF band	3–30 MHz
Very high frequency	VHF band	30–300 MHz
Long wave	L band	1–2 GHz
Short wave	S band	2–4 GHz
Compromise between S and X	C band	4–8 GHz
Crosshair	X band	8–12 GHz
Kurz-under	K_u band	12–18 GHz
German Kurz (short)	K band	18–27 GHz
Kurz-above	K_a band	27–40 GHz
V band	V band	40–75 GHz
W band	W band	75–110 GHz
Long wave	G band	110–300 GHz

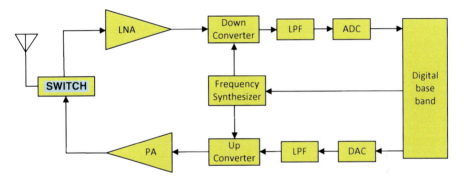

Fig. 1.1 Simple RF transceiver architecture [10]

consumption. The direct down-conversion architecture has become more popular nowadays. However, compared to the heterodyne architecture, it is easier to integrate the complete system on a single chip. Various technologies such as BiCMOS, SiGe HBT, and Bipolar are used to design radio-frequency (RF) switches; however, CMOS technology is very suitable for integration of both analog and digital circuits on a single chip. So the CMOS technology is preferred for implementation of RF front-end circuitry.

A basic heterodyne RF transceiver front-end system is shown in Fig. 1.1. In this architecture, the received RF signals are first passed through a band-pass filter and then switched to an low noise amplifier (LNA). Due to its gain, the LNA essentially sets the signal-to-noise ratio for the receiver chain. The amplified signals are filtered for improved image rejection and down-converted to an intermediate frequency (IF) with a mixer. The signals at IF are then filtered for channel selection and shifted in frequency to baseband by a second mixer [5–7]. However, the transmission process is complementary to the reception process. During the transmission, the signals at baseband are up-converted to the RF carrier using an IF stage. The power amplifier (PA) is used to drive the antenna. A transceiver (T/R) switch is used to connect/disconnect the antenna for transmit and receive processes. The direct down-conversion or homodyne architecture mixes the incoming RF signals with the carrier frequency to generate signals directly at baseband. Similarly, the signals are directly up-converted to the RF carrier using only one mixing step during transmission. The integrated circuit design industry is increasingly improving the direct down-conversion architectures to facilitate further integration by reducing the number of components required. This architecture uses standard CMOS technology and includes an LNA and PA on the same piece of silicon.

However, highly integrated transceiver solutions for the 802.11b/g standards have also been presented by Chien [8] and Kluge [9]. Among the component blocks of Fig. 1.1, the transceiver switch stands out as a candidate for on-chip integration because the MOSFET device is optimized to operate as a switch. In early days, the radio transceiver switches have been designed using PIN diode and FET, which consume more power. As the modern portable devices demand less power consumption switches, therefore the PIN diodes and FETs are gradually replaced by the MOSFETs such as n-type MOSFET and p-type MOSFET [10–12]. The MOSFET analog switches use the MOSFET channel as a low ON-state resistance switch to pass analog signals at switch-ON condition and as high impedance at switch-OFF condition and the signals flow in both directions across a MOSFET switch. The source is a negative side for n-type MOSFET or more positive side for p-type MOSFET. All of these switches are limited on what signals they can pass or stop by their gate–source voltage, gate–drain voltage, and source–drain voltages; exceeding the voltage, current, or power limits will potentially damage the switch.

In the latest technologies, CMOS fabrications have resulted in deep submicron transistors with higher transit frequencies and lower noise figures, so the trend started to explore the use of CMOS technology in RF circuits. This is also in the view of a system on a chip realization, where digital, mixed-signal baseband, and HF transceiver blocks would be integrated on a single chip. This technique has the ability to integrate RF circuits. Other advantages offered by silicon CMOS technologies are the low cost due to the volume of wafers processed and the low power consumption feature of MOSFETs, which makes it suitable for portable applications. It has been known that for analog and RF applications, the accuracy of circuit simulations is strongly determined by the device models. However, the accurate device models become crucial to correctly predict the circuit performance.

The RF/microwave switching elements using silicon CMOS technology are being investigated and presented as an alternative to the traditional PIN diode and GaAs MESFET devices. The silicon CMOS RF switching elements are attractive because of their potential application in all silicon monolithic CMOS solutions for completely integrated baseband and RF functions in low-cost wireless, mobile satellite, and personal communication systems. RF switches can be used at several places in RF front ends. In a transceiver switch, a double-pole single-throw (DPST) arrangement of switches multiplexes the use of the antenna between the PA and LNA. The transceiver switches must have a high linearity to ensure that the high-power signals (~2 W) at the output of the PA are transmitted to the antenna with minimum distortion [13, 14]. This linearity requirement presents a challenge to integrate transceiver switches into on-chip designs especially as the supply voltage in standard CMOS continues to decrease.

In addition to the transceiver switch application mentioned above, RF switches could be used to select capacitors, for example, tuning of a voltage-controlled oscillator (VCO). In this application, the potential challenge is to obtain a low ON-state resistance and a low OFF-state capacitance. In a given technology, the ON-state resistance and OFF-state capacitance are inversely related to each other since the resistance–capacitance product in a modern CMOS technology is not as low as desired [15, 16].

In the receiver part of the communication system (a handset), low-noise RF transistors are used to amplify the incoming signals. As in any LNA, the use of low or minimum noise figure transistors is desired. The noise requirements for the RF devices for this application are, however, not as stringent as those for the satellite communications. In wireless communications, the receiver experiences the noise of the environment, which is interference-dominated, whereas in the satellite communications, the signal comes from the sky with less background noise [17].

Consequently, for the wireless communications, the noise produced intrinsically in the RF devices is somewhat negligible comparing to that from the noisy environment. However, another requirement for the communication system is the reduction of power consumption. At present, a supply voltage of 3 V has been established as a standard [18]. To deliver a high output power combined with a high efficiency at a limited supply voltage of 3 V, RF power transistors possessing a large ON-state current and a low ON-state resistance are required in the transmit section of the handset.

1.2 Radio-Frequency Switches

In the radio transceiver of the advanced communication systems, multiple antenna system is used to replace the traditional single antenna circuitry to improve the transmission capability and reliability. In the antenna selection system, the signals from a subset of the antennas are processed at any time by the limited bandwidth of RF, which is available for the receiver. Hence, the transmitter needs to send pilots multiple times to enable the receiver to estimate the channel state of all the antennas and select the best subset. With the multiple antennas, the data transfer rate can be increased by the same factor. For example, if we have n antennas as $a_1, a_2, a_3, \ldots, a_n$ used in the transceiver, then data transfer rate will increased by factor of n as it is number of antenna used. For such communication system, the antenna selection and switch mechanism is essential to circumvent the uses of several RF chains, associated with the various antennas. The desired switching system must have a simple and low-cost structure which also confined all the improvement of multiple-input, multiple-output (MIMO) systems [19, 20].

A traditional n-type single-pole double-throw (SPDT) MOSFET switch has good performance but only for a single operating frequency. For multiple operating frequencies, the high signal distortions are easily observed, which results in an unrecognizable information signal at the receiver end. To be able to transmit or receive information through the multiple antenna systems, known as MIMO systems, it becomes necessary to design a new RF switch that is capable of operating with multiple antennas and frequencies as well as minimizing signal distortions and power consumption [21, 22]. For this purpose, we have analyzed a model of double-pole four-throw (DP4T) RF CMOS switch and achieved a better performance with respect to drain current, switching speed, and the voltages as compared to the single-pole double-throw (SPDT) [23], double-pole double-throw (DPDT) [24], and single-pole four-throw (SP4T) [25] switches. This explored switch is low in cost and capable of selecting data streams to or from the two antennas for transmitting or receiving processes, respectively.

We have started with a basic CMOS inverter switch and reached up to double-pole four-throw double-gate (DP4T DG) RF CMOS switch. It is designed with low insertion loss and low control voltage. The advantages of this CMOS inverter switch are its minimal distortion and negligible voltage fluctuation and do not require large resistance at the receiving end. In addition to this, DP4T switches can be easily implemented into MIMO systems to increase the diversity and system capacity due to the multiple antenna usage. We have designed a novel cylindrical surrounding double-gate (CSDG) MOSFET. After that we have designed these MOSFETs with high dielectric material and analyzed its property. To validate the design correctness (surface smoothness), we also design a model using image processing.

Huang [25] has fabricated an SPDT transceiver switch [26] for 3.0 V for a 0.5-μm CMOS process. The discussed analysis in [25] shows that the substrate resistances and source/drain-to-body capacitances must be lowered to decrease insertion loss. This switch exhibits a 0.7-dB insertion loss, 17-dBm compression point (P_1 dB), and 42-dB isolation at 928 MHz. The switch has adequate insertion loss, isolation, P_1 dB, and IP_3 for 900 MHz. The ISM band applications require a moderate peak transmitter power level (~15 dBm). To avoid the uses of multiple RF chain associated with the multiple antennas (used to replace traditional single antenna circuitry in the radio transceiver system in order to improve the transmission capability and reliability) and RF switch is the most essential component. Mekanand and Eungdamorang [27] have proposed that the DP4T RF CMOS switch at frequencies 2.4 GHz and 5.0 GHz exhibits an insertion loss of 0.75 dB and 0.86 dB, respectively, with compression point of 31.86 dBm and realizes the minimal distortion, negligible voltage fluctuation, and low power supply of only 1.2 V, which is used in wireless local area network (WLAN) and other advanced wireless communication systems. They also discussed the advantages of switch using a CMOS, instead of a single n-MOS switch in the dynamic range. This dynamic range in the ON-state is significantly increased, which allows a full signal swing. Moldovan et al. [28] have demonstrated the analytically compact undoped DG MOSFET model and forecast the effect of volume inversion on the intrinsic capacitances and reveals that the transition from volume inversion regime to the double-gate behavior. This result shows that the intrinsic capacitances are larger as well as limit the high-speed (delay time) behavior of the DG MOSFETs under volume inversion regime. Lee et al. [29] have presented a novel architecture for the DPDT, DP4T, and 4P4T RF switches with simple control logics and high-power handling capabilities, which require only one, two, and three control lines, respectively. However, the developed DPDT switch demonstrates 1.0 dB of insertion loss, 19 dB of isolation, and 31 dBm of input P_1 dB, 34.5 dBm of input P_1 dB in 3/0 V operation at 5.8 GHz. The DP4T and 4P4T switches exhibit 1.8 dB, 2.8 dB insertion loss, and 23/37 dB, 20/35/55 dB of isolation, respectively, and 31 dBm of input P_1 dB, 35 dBm of input P_1 dB in 3/0 V operation at 5.8 GHz.

Woerlee et al. [30] have presented the impact of scaling on the analog performance of MOS devices at radio frequencies and explored the trends in the RF performance of nominal gate length of n-MOS devices from 350- to 50-nm CMOS technologies. The RF performance metrics such as the cutoff frequency, maximum oscillation frequency, power gain, noise figure, linearity, and $1/f$ noise have been explored in the analysis [31]. The minimum gate length of n-type MOSFET devices as 350-, 250-, and 180-nm CMOS technologies was studied.

Lee [32] has presented the standard digital CMOS process, which offers a number of ways to improve the characteristics of on-chip passive elements. In particular, it is possible to reduce significantly the severity of substrate loss. It is clear from [33] that the scaling trends properly exploited and combined with new insights into the device and oscillator noise enable the CMOS IC technology to perform at GHz frequencies to make it attractive for application specific once thought the sole province of more exotic technologies.

These advantages include high integration capabilities and excellent electrostatic discharge (ESD) robustness. However, some RF switches do not require external DC blocking capacitors and have the control logic fully integrated. With CMOS compatible logic levels, the need for external level shifters is eliminated, as there are no any external components are required. So, the RF switches enable the really tiny board designs.

1.3 Radio-Frequency MOSFETs

Earlier approaches to RF MOSFET modeling involve adding lump elements to a compact model for digital and analog circuit designs, such as BSIM3, BSIM4, and MM9, and focused on how to build a reasonable sub-circuit and how to extract their values according to the equivalent circuits. These methods consist of the analysis and optimization. However, few attempts have been made to build a scalable RF MOSFET model, including the layout-based extrinsic elements. The RF MOSFETs are the MOSFETs that are designed to handle high-power RF signals from devices such as stereo amplifiers, radio transmitters, and television monitors. For the RF MOSFET modeling, various issues need to be considered, especially the three most important parasitic components: gate resistance, input impedance, and noise performance.

The RF MOSFETs are turned ON and turned OFF by input voltages and function as miniature electronic switches. However, such as other semiconductor devices, the RF MOSFETs are made of materials like Silicon (Si) or Germanium (Ge) and doped with impurities to induce changes in the electrical properties. The voltage is applied between the gate and source terminals, which modulate the current between the source and drain. Typically, a thin layer of oxide insulation is used to prevent the current from flowing between the gate and a conductive channel in the semiconductor substrate [34].

The RF MOSFETs also seem to fulfill most of the performance requirements for RF systems with operating frequencies up to 6 GHz [35, 36]. The RF CMOS transistors are combination of n-channel MOSFET and p-channel MOSFET. With both types of devices, the polarity of the electric field that controls the current in the channel is determined by the majority of carriers in the channel. The selection of RF MOSFETs requires an analysis of performance specifications, such as:

a. Drain–source breakdown voltage is the maximum drain-to-source voltage before the breakdown with the gate grounded.
b. Power gain in dB is a measure of the power amplification, which is the ratio of the output power to input power.
c. Noise figure in dB is a measure of the amount of noise added during normal operation, which is the ratio of the signal-to-noise ratio at the input and the signal-to-noise ratio at the output.
d. Power dissipation in W or mW is a measure of total power consumption.
e. Attenuation is measure of loss of signal in the transmission through MOSFET.
f. Gain in dB is a measure of the ratio of output quality to the input quality.
g. Switching speed means how fast switch ONs and OFFs.

However, some other performance specifications for RF MOSFETs include the maximum drain saturation, common-source forward transconductance, operating frequency, and output power. The devices that operate in depletion mode can increase or decrease their channels by an appropriate gate voltage. By contrast, the devices that operate in enhancement mode can only increase their channels by an appropriate gate voltage. Some bipolar RF MOSFETs are suitable for automotive, commercial, or general industrial applications. The RF MOSFETs vary in terms of operating mode, packaging, and packing methods.

The requirements of a MOSFET model in RF application such as continuity, accuracy, and scalability are necessary. In addition to these properties, some other important requirements to the RF models are given below [37]:

a. Predict bias dependence of small-signal parameters at RF operation.
b. Describe the nonlinear behavior of the devices in order to permit accurate simulation of intermodulation distortion and high-speed large signal operation.
c. Predict RF noise, which is important for the design of LNAs.
d. The components in the developed model should be physics based and geometrically scalable so that the model can be used in predictive and statistical modeling for RF applications.

Moreover, the model should be derived with the inclusions of normal and reverse short-channel and narrow-width effects, velocity saturation, channel length modulation, polysilicon depletion, velocity overshoot, drain-induced barrier lowering, mobility degradation due to vertical electric field, impact ionization, band-to-band tunneling, self-heating, and channel quantization. Various MOSFET models, including MOS9, EKV, and BSIM3v3 [38], have been developed for the digital, analog, and mixed-signal applications. Recently, they all are extended for use in the RF applications. In the continuation with these developments, we have proposed a model of double-gate MOSFET and then extend it for the cylindrical surrounding double-gate (CSDG) MOSFET and use a high dielectric material (e.g., hafnium dioxide) at some extent.

1.4 Issues of Radio-Frequency MOSFET Modeling

At low frequencies, the impedance of the junction capacitance is so large that the substrate impedance may not be seen from the drain terminal, and a MOSFET can be modeled as a three-terminal device [39]. This three-terminal device is treated as a two-port network, where four complex numbers are sufficient to characterize the device. On the other hand, nine complex numbers are required to characterize a three-port network, such as a four-terminal device. Even if the three-port measurements are possible, many bias combinations for the AC characteristics should need to be considered, and the measurements and parameter extractions become impractical and time consuming. In the MOSFET, the number of free carriers available in the channel is mostly controlled by the field induced from the gate voltage; however, a change in the body potential can also affect the number of channel carriers. In spite of these problems, intensive efforts have been devoted to RF MOS modeling and parameter extraction.

A traditional n-MOS switch uses n-MOS as transistors in its main architecture and requires the control voltage of 5.0 V and a large resistance at the receivers; consequently, antennas will detect the signal. The CMOS IC switch is an integrated circuit using FETs to achieve switching between the multiple paths, because of its high value of control voltage. However, it is not suitable for the modern portable devices which demand the smaller power consumption.

The aggressive scaling of MOSFET has led to the fabrication of high-performance MOSFETs with a cutoff frequency more than 150 GHz [40]. As a result of this development, the CMOS is a strong candidate for the RF wireless communications at GHz frequency range. However, accurate device models are needed to design the advanced RF circuits and in this regard, various researcher proposed some parameters for cell design as used for RF switch circuits [21–25, 41]. The large cellular systems can now be processed on a single chip containing most of the transceiver blocks at RF and baseband. To achieve the performance of GaAs switches, the CMOS-based RF switches had to be manufactured on dedicated and more expensive sapphire wafers. The new Infineon's RF switches combined the benefits of CMOS with outstanding RF performance. The switches have RF performance traits such as low insertion loss, low harmonic distortion, better isolation, and high power levels. The RF switches also have the inherent CMOS advantages including high integration capabilities, cost effectiveness, and excellent electrostatic discharge (ESD) robustness.

The layout of a MOSFET is usually drawn in the micron rules of the technology. When a new technology introduces, then the layout of any device can be migrated. There may be a predicament that if the original layout has used all the minimum widths and spacing which are then incompatible with the rules of the new technology can be used. To solve these problems, Mead and Conway [16] have proposed a model to draw the layout in a nominal 2-μm layout and then apply a λ-scaling factor to the desired technology because this scaling factor is based on the pitch of various elements like transistors, metal, and poly.

Over the past decades, the MOSFET has continually been scaled down in size. The modern integrated circuits are incorporating MOSFETs with channel lengths of tens of nanometers. The semiconductor industry maintains a roadmap, the International Technology Roadmap for Semiconductors (ITRS) [34], which sets the pace for MOSFET development. The ITRS is sponsored by the five leading chip manufacturing regions in the world, such as Korea (5 %), Europe (10 %), Taiwan (13 %), Japan (17 %), and the United States (55 %). The difficulties with decreasing the size of the MOSFET have been associated with the semiconductor device fabrication process, the need to use very low voltages, and better electrical performance necessitating circuit redesign and innovation. The main reason to make transistors smaller is to pack various devices in a smaller single chip. Since the fabrication costs for a semiconductor wafer are relatively fixed, the cost per integrated circuits is mainly related to the number of chips that can be produced per wafer. Hence, the integrated circuits (ICs) allow more chips per wafer, reducing the price per chip. In fact, over the past 30 years the number of transistors per chip has been doubled every 2–3 years once a new technology node is introduced. For example, the number of MOSFETs in a microprocessor fabricated in a 45-nm technology can well be twice as many as in a 65-nm chip. This doubling of transistor density is commonly referred to as Moore's law [16].

However, the continuous scaling of CMOS technology has now reached a state of evolution, in terms of both frequency and noise, where it is becoming a severe part for RF applications in the GHz frequency range. With the scaling of device dimensions and increase in the short-channel effects (SCE), the multiple-gate transistors have been investigated to obtain an improved gate control characterization [42–44]. Due to these advantages, there has been growing interest in modeling of RF CMOS which is especially striking for various applications because it allows integration of both digital and analog functionality on the same die, increasing performance at the same time as keeping system sizes reserved [45–48]. An excellent improvement in the frequency response of Si-CMOS devices has explored their application in the microwave/millimeter wave regime of the electromagnetic spectrum for various wireless communication systems such as high-capacity

1.4 Issues of Radio-Frequency MOSFET Modeling

wireless local area network, short-range high data rate, wireless personal area networks, and collision avoidance radar for automobiles.

For the investigation of circuit-level degradation, a CMOS inverter is analyzed [49–51]. A major advantage of CMOS technology is the ability to combine complementary transistors, n-channel and p-channel, on a single substrate. Recently, the CMOS transistor uses the technique of silicon on insulator (SOI), which is very attractive because of the high-speed performance, low power consumption, its scalability, and effective potential [52, 53]. As compared to the bulk silicon substrate, the architecture of SOI MOSFET is more flexible due to several parameters such as thicknesses of film and buried oxide, substrate doping, and back-gate bias which are used for optimization and scaling. The short-channel effect, junction capacitances, and doping fluctuation are mitigated in ultrathin SOI films [54, 55]. The main advantage of SOI as compared to bulk silicon is its compatibility with the use of high resistivity substrates to reduce the substrate coupling and RF losses [56]. Saha [57] has solved a problem of critical issue with the continued scaling of MOSFET devices towards their ultimate dimensions near 10-nm regime and defined the process variability; induced device performance variability has become a critical issue in the design of very large-scale integrated (VLSI) circuits using advanced CMOS technologies. Manku [58] has discussed the design issues and the microwave and RF properties of CMOS devices and a qualitative understanding of the microwave characteristics of MOS transistors is provided. This design is helpful for integrated circuit design to create better front-end RF CMOS circuits and presented the network properties of CMOS devices, frequency response, microwave noise properties, and scaling rule. Srivastava et al. [59] have designed a model for a capacitor device using MOS layers with an oxide thickness of 528 Å (measured optically) and measured the material parameters from the curve drawn between capacitance versus voltage and capacitance versus frequency. In this work, to find better result, Srivastava et al. [59] varied the voltage with smaller increments and performed the measurements. Srivastava et al. [60, 61] have presented a model for a capacitor device, and measured the oxide thickness optically. Its accuracy depends on the quality of models, parameters, and numerical techniques it employs and also validates the result by measurement of capacitance at different voltages; using LCR meter based on this oxide thickness measurement of a MOS capacitor, one can measure the device parameters, mainly the substrate dopant concentration.

Ye and Cao [62] have realized that the random variations have been regarded as one of the major barriers on CMOS scaling and proposed a compact 3-D model and perform the atomistic simulations to investigate the fundamental variations in a scaled CMOS device, including random dopant fluctuation, line edge roughness, and oxide thickness fluctuation. These models are scalable with device specifications, enabling quantitative analysis of circuit performance variability in the future technology nodes. Caverly et al. [63] have fabricated some cells using standard processes of 2.0-, 1.2-, and 0.8-µm CMOS integrated circuit without postprocessing performed. The results indicate that 2.0-µm CMOS can be used successfully up to approximately 0.25 GHz with 0.8-µm cells useful up to approximately 1 GHz. Srivastava et al. [64] have investigated the design parameters of RF CMOS cells which are suitable for switch in the wireless telecommunication systems, and this results for the development of a cell library which includes the basics of the circuit elements required for the radio-frequency subsystems of the integrated circuits such as voltage–current (V–I) characteristics at low voltages, contact resistance which is present in the switches, and the potential barrier with contacts available in the devices. The rapid integrated system designs are the use of cell libraries for various system functions [65, 66]. In digital system design, these standard cells are both at the logic primitive level and higher levels of circuit functionality. For baseband analog systems, the standard cell libraries are less frequently used. In the design of a CMOS RF cell library, the cells must be designed to be flexible in terms of drive requirements, the bandwidth, and circuit loading. For RF applications, the widespread drive requirements for off-chip loads are based on 50 Ω impedances. This impedance is a good compromise between lowest loss and highest power handling for a given cable size. Also, this impedance caught on for RF transmission rather than the well-established 75 Ω that had been used for video transmission. A factor governing the bandwidth of the RF cells is the nodal capacitance to ground, primarily the drain and source sidewall capacitances [67, 68]. Since these cells are to be used with digital and baseband analog systems, control by on-chip digital and analog signals is another factor in the design [69].

The library consists of cells design, using standard 2.0- and 0.8-μm CMOS processes. For the technologies studied, these control voltages varied between 0.0 and 2.1 V, with the supply voltage of 1.2 V, are of interest for low power consumption portable system applications. The cells have been designed for the purpose of radio-frequency communication switch devices and high-power RF MOSFET targets VHF applications. The transistors designed for the purpose of library elements are usually planned with multiple gate fingers to reduce the capacitances of sidewall. This increases the contact resistance and reduces the barrier height. The properties for RF CMOS switch design for the application in communication [70] and designed results are presented in Chap. 2 and have been designed with and optimized for the particular application. However, higher drain current can be easily achieved by using larger number of gate finger which is also analyzed.

However, the 5.0 GHz and higher frequencies are important because they include several commercial communication bands. Although the 2.0-μm cells had a lower operating frequency range than the smaller process geometries, this technology is still useful for several applications such as the processing of RF signals in the 250 MHz range. The design goals were met for all the cell library elements such as RF control elements, single-ended class-A amplifier, RF isolator, and Gilbert cell mixer circuit. In cell libraries of the RF MOSFET, designed cells should have elasticity for working frequency range, driving capability, and loading of devices or circuits. For the RF applications, generally we used off-chip load impedance, which is of 50 Ω [71, 72]. RF MOSFET has features of best noise figure and gain, excellent cross-modulation, ESD robustness 5th generation MOSFETs. For example

BF5030W, BG5120K fulfill the stringent technical requirement from the digital tuners, and as well support low-power 3 V designs, automotive quality [73].

The performance of a transceiver switch is characterized by several parameters in transmit and receive modes such as insertion loss, isolation, return loss, linearity, power, ESD reliability, ON-state resistance, and capacitances [74, 75]. For optimizing the performance of a MOSFET as an analog switch requires number of trade-offs. If the width-to-length ratio is increased to reduce R_{ON}, the parasitic capacitance of the gate oxide increases proportionately and provides lower bandwidth. There are many factors that affect R_{ON} such as temperature, input voltage, supply voltage, and gate length. The R_{ON} is application specific; however, for low-signal, high-speed applications, small R_{ON} is required to maintain the integrity of the input signal throughout the device. If the application is for the audio, a low R_{ON} may not be as important due to the lower frequencies and power of the signal.

If a MOSFET is being used for a power application as the source-to-drain voltage increases, a larger source-to-drain breakdown voltage is required, therefore increasing the value of R_{ON}. In order to design a switch or any type of MOSFET device, a complete understanding of the application is required. In the p-type MOSFET gate region, an increase in the parasitic capacitance is a concern at high frequencies. In this situation, the gate oxide becomes an issue especially at high frequencies. The source-to-substrate capacitance, the drain-to-substrate capacitance, and the source-to-drain capacitance have an effect on R_{ON}. As the frequency of the input signal increases, these capacitances can increase insertion loss and decrease off-isolation. With R_{ON} being nonlinear and a strong function of voltage and temperature, an analog switch should not be used in any critical resistive voltage divider path in a circuit. The ESD performance of the switch is usually measured using the human body model (HBM) method. This method essentially measures the robustness of the part when subjected to a static discharge arising from human contact. With this model, a 200 pF capacitor is charged to a certain voltage and then discharged through the device under test (DUT). The DUT breaks down and ceases to function as the voltage on the capacitor exceeds a certain threshold. This breakdown voltage is used as a measure of the ESD reliability of the DUT. However, most of the GaAs RF components are rated at 500 V HBM and other practical requirements of the transceiver switch include robustness with respect to antenna mismatch. Also, the control voltage levels used to toggle the state of the switch must be available in the system. The turn-ON and turn-OFF times are typically less than 10 ns to enable rapid transition between T_x and R_x modes, although specific values depend on the application [76].

1.5 Double-Gate MOSFET

The figure of merit for MOS device switching speed increases as the inverse square of the MOS transistor physical channel length, which encourages the shrinking of the device size. The basic idea is that the output or drive current of a device available to switch its load devices increases linearly as its channel length decreases, while the current required by the load devices to achieve switching decreases as their gate area and hence their channel length decreases. There is a twofold benefit in device dimension reduction [77]. Since the drive current requirement to switch the load devices depends upon the total load capacitance and area, there is a strong motivation to reduce the size of the complete device, not only its channel length.

The progress to scale down the transistors to smaller dimensions provides faster transistors as well as lowers the effective costs per transistor and density in terms of transistors area. The transistor scaling necessitates the integration of new device structures. The double-gate (DG) MOSFETs are example of this, which are capable for nanoscale integrated circuits due to their enhanced scalability compared to bulk or Si-CMOS [78, 79]. A variety of multiple-gate nonclassical CMOS structures have been proposed in [80] and demonstrated to manage electrostatic integrity (i.e., short-channel effects) in ultra-scaled CMOS structures. In the first of these structures, the N-gate ($N > 2$) MOSFET [81, 82], current flows horizontally (parallel to the plane of the substrate) between the source and drain along vertical channel surfaces, as well as one or more horizontal channel surfaces. The large number of gates provides for improved electrostatic control of the channel, so that the Silicon body thickness and width can be larger than for the ultrathin body SOI and double-gate FET structures, respectively.

However, various double-gate MOSFET structures have been proposed to further improve the engineering of the channel electrostatics and to provide independent control of two isolated gates for low-power and mixed-signal applications. In the *tied double-gate, sidewall conduction structure* [83, 84], current flows horizontally (parallel to the plane of the substrate) between the source and drain, along the opposite vertical channel surfaces. The width of vertical silicon fin is narrow (smaller than the channel length) to provide adequate control of short-channel effects. The principal advantage with this structure is the planar bulk-like layout and process. In the *tied double-gate planar structure* [85, 86], the current flows horizontally (parallel to the plane of the substrate) between the source and drain along opposite horizontal channel surfaces. The principal advantages of this structure reside in the potential simplicity of the process and in the compactness of the layout as well as in its compatibility with bulk layout. However, another major challenge is in the fabrication process, particularly for those structures requiring alignment of the top- and bottom-gate electrodes. The *independently switched double-gate (ground-plane) structure* [87, 88] is similar to the tied double-gate planar structure, except that the top- and bottom-gate electrodes are electrically isolated to provide for independent biasing of the two gates. The top gate is typically used to switch the transistor ON and OFF, while the bottom gate is used for dynamic (or static) V_{th} adjustment. In the *vertical conduction structures* [89, 90], the current flows between the source and drain in the vertical direction along two or more vertical channel surfaces. The main advantage of this structure is that the channel length is defined by epitaxy rather than by lithography. This structure requires a challenging fabrication process which is a disadvantage for this design.

The symmetric DG MOSFET structure focuses on the application of RF switches due to its intrinsic strength to the short-channel effects and improves current drive capability. When we are using a switch with multiple gates, the behavior of these switches depends on the number of gates, which controls the operational process of the device. So the additional logic functions can be implemented into a single-chip transistor. The transistors that use independently controlled gates are not limited to only two gates, but for the geometrical reasons of the transistor and the connectivity of the transistor terminals, it is suitable to use only two gates. The independent double-gate transistors can be used to implement the universal logic functionality within a single transistor [90]. Gidon [91] has investigated the two-dimensional DG MOSFET and combated the high aspect ratio of the transistor (thin channel compared to its length) by introducing an anisotropy scale factor in its geometry. Taur and Ning [12] have presented an analytic potential model for the long-channel symmetric and asymmetric DG MOSFETs. The model is derived rigorously from the solution to the Poisson's continuity equation and current continuity equation without the charge-sheet approximation.

To preserve the proper physics, volume inversion in the subthreshold region is well accounted for the model. For these analytical expressions of the drain current, terminal charges and capacitances for the long-channel DG MOSFETs are obtained continuously in all operation regions, such as linear, saturation, and subthreshold. The drain current model, charge model, transconductance model, and capacitive model for symmetrical and asymmetrical DG MOSFETs are also developed in [88].

1.6 Cylindrical Surrounding Double-Gate MOSFET

We have analyzed the design parameters of cylindrical surrounding double-gate (CSDG) MOSFETs for the advanced wireless telecommunication systems. The proposed cylindrical surrounding double-gate radio-frequency complementary metal-oxide-semiconductor (CSDG RF CMOS) device is operating at the microwave frequency regime of the spectrum. This MOSFET can be used as the RF switch for selecting the data streams from antennas for both the transmitting and receiving processes [45, 92–94]. We have emphasized on the basics of the circuit elements (e.g., drain current, threshold voltage, resonant frequency, resistances at switch-ON condition, capacitances, energy stored, cross talk, and switching speed) required for the integrated circuit of the radio-frequency subsystem of the CSDG RF CMOS device and role of these basic circuit elements is also discussed [95–97]. However, various properties demonstrated by the switch, due to the cylindrical surrounding double-gate MOSFET, have been discussed in this book.

We have presented a model of the low-power and high-speed RF CMOS with CSDG structure. The layout of the design has been studied to understand the effect of device geometry. Each of the parameters is discussed separately for the purpose of clarity of presentation and understanding the operation of CSDG RF CMOS switch structures [98].

1.7 Hafnium Dioxide-Based MOSFET

The reduction in sizing ratio of the gate dielectric, which works as a capacitor in the MOSFET, results in the increase of capacitance and speed of the device. However, this process has reached up to the limit when the further reduction of the SiO_2 thickness results in the leakage current above the acceptable limit [99, 101]. This problem can be resolved by replacing SiO_2 with materials having higher dielectric constants. Hafnium dioxide (also known as Hafnia) is one of them, which has relatively large energy band gap and a better thermal stability as compared to silicon [102–104]. The gate materials should be thermodynamically stable on the gate dielectric and must be able to withstand high temperature used in the fabrication of device. In device fabrication process for the activation of dopant atoms in the source, drain, and gate regions of the transistor, highest temperatures are used. Hence, the new gate electrode material must be chosen with high-k dielectric material. The motivation to replace traditional SiO_2 gate dielectrics with HfO_2 is because it allows increased gate capacitance without affecting the leakage effects [105, 106]. In order to improve the performance of MOSFET devices, HfO_2-based gate layers are being integrated to achieve low leakage current.

1.8 Image Acquisition of the MOSFETs

The image acquisition is frequently used in systems for monitoring and controlling of objects to support effective management of their resources [107–109]. The practical systems for monitoring rectangular objects, like DG MOSFET, and cylindrical objects, like CSDG MOSFET, have been discussed in Chap. 7. This system requires various vision sensors, recording images that have to be transmitted to and processed in the central processing unit [110–113]. One of the most challenging problems in such cases is effective transmission and processing of huge amount of image data. To avoid overloading of transmission channels and a central unit, various already existing algorithms are frequently performed at the sensors by an integrated low-level image processor. A complete vision chip consisting of a photodetector array can effectively be implemented on DG MOSFET and CSDG MOSFET, which are formed on the rectangular and cylindrical substrate, respectively [114, 115].

1.9 Conclusion

From the above discussions, we conclude that to transmit and receive the data with twice the speed for communication systems, the design of DP4T RF switch is a good idea [116]. So, we designed a DP4T RF CMOS switch and do some modification in this switch with the replacement of MOSFET with DG MOSFET. After the DG MOSFET, we have designed the CSDG MOSFET and then replace the substrate from SiO_2 to HfO_2. Finally, we design a model to check the correctness of the design structure.

References

1. William Fothergill Cooke, *The Electric Telegraph*, W. H. Smitha and Sons, London, 1857.
2. Hubbard and Geoffrey, *Cooke and Wheatstone and the Invention of the Electric Telegraph*, Routledge and Kegan Paul Publications, London, 1965.
3. Christopher Buff, "Radio receivers-past and present," *Proc. of the IRE*, vol. 50, no. 5, pp. 884–891, May 1962.
4. U. S. Bureau of Naval Personnel, *Basic Electronics*, Courier Dover Publications, 1973.
5. M. Steyaert J. Janssens, B. De Muer, M. Borremans, and N. Itoh, "A 2 V CMOS cellular transceiver front-end," *IEEE J. of Solid State Circuits*, vol. 35, no. 12, pp. 1895–1907, Dec. 2000.
6. Behzad Razavi, "Challenges in portable RF transceiver design," *IEEE Circuits and Devices Magazine*, vol. 12, no. 5, pp. 12–25, Sept. 1996.
7. G. Boeck, D. Pienkowski, R. Circa and R. Kakerow, "RF front-end technology for reconfigurable mobile systems," *Proc. of IEEE Int. Conf. on Microwave and Optoelectronics*, Brazil, 20–23 Sept. 2003, pp. 863–868.
8. G. Chien, F. Weishi, Y. Hsu, and L. Tse, "A 2.4 GHz CMOS transceiver and baseband processor chipset for 802.11b wireless LAN application," *Proc. of IEEE Int. Conf. on Solid State Circuits*, San Francisco, USA, 9–13 Feb. 2003, pp. 358–359.
9. W. Kluge, L. Dathe, R. Jaehne, and D. Eggert, "A 2.4 GHz CMOS transceiver for 802.11b wireless LANs," *Proc. of IEEE Solid State Circuits Conf.*, San Francisco, USA, 9–13 Feb. 2003, pp. 360–361.
10. Behzad Razavi, *RF Microelectronics*, 3rd Edition, Prentice Hall, New Jersey, 1998.
11. Application note, *Design with PIN Diodes*, APN 1002, Skyworks Solutions Inc., Woburn, MA, July 2005.
12. Yuan Taur and Tak Ning, *Fundamentals of Modern VLSI devices*, Cambridge University Press, USA, 2009.
13. J. P. Carmo, P. M. Mendes, C. Couto, and J. H. Correia, "A 2.4 GHz RF CMOS transceiver for wireless sensor applications," *Proc. of Int. Conf. on Electrical Engineering*, Coimbra, Portugal, 1–5 Oct. 2005, pp. 902–905.
14. Annapurna Das and Sisir K. Das, *Microwave Engineering*, 2nd Edition, Tata McGraw-Hill, India, 2009.
15. R. Granzner, S. Thiele, and Frank Schwierz, "Quantum effects on the gate capacitance of tri-gate SOI MOSFETs," *IEEE Trans. on Electron Devices*, vol. 57, no. 12, pp. 3231 – 3238, Dec. 2010.
16. Kaushik Roy and Sharat C. Prasad, *Low Power CMOS VLSI Circuit Design*, 1st Edition, Wiley, India, 2009.
17. N. Wang, "Transistor technologies for RFICs in wireless applications," *Microwave J.*, pp. 98–110, Feb. 1998.
18. Jack Browne, "More power per transistor translates into smaller amplifiers," *Microwaves and RF*, vol. 6, pp. 132–136, Jan. 2001.
19. Peerapong Uthansakul, Nattaphat Promsuwanna, and Monthippa Uthansakul, "Performance of antenna selection in MIMO system using channel reciprocity with measured data," *Int. J. of Antennas and Propagation*, vol. 2011, pp. 1–10, 2011.

20. S. Sanayei and N. Aria, "Antenna selection in MIMO systems," *IEEE Communications Magazine*, vol. 42, no. 10, pp. 68–73, Oct. 2004.
21. A. M. Street, *RF Switch Design*, IEE Training Course, United Kingdom, vol. 4, pp. 1–7, April 2000.
22. Kevin Walsh, *RF Switches Guide Signals in Smart Phones*, Skyworks Solutions Inc., Woburn, MA, Sept. 2010,
23. Qiuting Huang, Paolo Orsatti, and Francesco Piazza, "GSM transceiver front-end circuits in 0.25 µm CMOS," *IEEE J. of Solid State Circuits*, vol. 34, no. 3, pp. 292-303, March 1999.
24. Zeji Gu, Dave Johnson, Steven Belletete, and Dave Frjklund, "A high power DPDT MMIC switch for broadband wireless applications," *Proc. of IEEE Radio Frequency Integrated Circuits Symposium*, Pennsylvania, USA, 8–10 June 2003, pp. 687–690.
25. F. J. Huang and O. Kenneth, "A 0.5 µm CMOS T/R switch for 900 MHz wireless applications," *IEEE J. of Solid State Circuits*, vol. 36, no. 3, pp. 486–492, March 2001.
26. K. Miyatbuji and D. Ueda, "A GaAs high power RF single-pole dual-throw switch IC for digital mobile communication system," *IEEE J. of Solid State Circuits*, vol. 30, no. 9, pp. 979–983, Sept. 1995.
27. Piya Mekanand and Duangrat Eungdamorang, "DP4T CMOS switch in a transceiver of MIMO system," *Proc. of 11th IEEE Int. Conf. of Advanced Communication Technology*, Korea, 15–18 Feb. 2009, pp. 472–474.
28. Oana Moldovan, Ferney A. Chaves, David Jimenez, Jean P. Raskin, and Benjamin Iniguez, "Accurate prediction of the volume inversion impact on undoped double-gate MOSFET capacitances," *Int. J. of Numerical Modeling: Electronic Networks, Devices and Fields*, vol. 23, no. 6, pp. 447–457, Nov. 2010.
29. C. Lee, B. Banerjee, and J. Laskar, "Novel T/R switch architectures for MIMO applications," *Proc. of IEEE MTT-S Microwave Symp. Digest*, 6–11 June 2004, vol. 2, pp. 1137–1140.
30. P. H. Woerlee, M. J. Knitel, and A. J. Scholten, "RF CMOS performance trends," *IEEE Trans. on Electron Devices*, vol. 48, no. 8, pp. 1776–1782, Aug. 2001.
31. N. Ashraf and D. Vasileska, "1/f Noise: threshold voltage and ON-current fluctuations in 45-nm device technology due to charged random traps," *J. of Computational Electronics*, vol. 9, no. 3–4, pp. 128–134, Oct. 2010
32. Thomas H. Lee, "CMOS RF no longer an oxymoron," *Proc. of the 19th Gallium Arsenide Integrated Circuit Symposium*, California, USA, 15–17 Oct. 1997, pp. 244–247.
33. S. Ahmed, C. Ringhofer, and D. Vasileska, "An effective potential approach to modeling 25 nm MOSFET devices," *J. of Computational Electronics*, vol. 9, no. 3-4, pp. 197–200, Oct. 2010
34. Joseph J. Carr, *Secrets of RF Circuit Designs*, 3rd Edition, Tata McGraw-Hill, India, 2004.
35. R. Langevelde and F. Klaassen, "An explicit surface potential based MOSFET model for circuit simulation," *Solid State Electronics*, vol. 44, no. 3, pp. 409–418, March 2000.
36. X. Xi, K. Cao, X. Jin, H. Wan, M. Chan, and C. Hu, "Distortion simulation of 90 nm n-MOSFET for RF applications," *Proc. of 6th Int. Conf. on Solid State and Integrated Circuit Technology*, Shanghai, China, 22–25 Oct. 2001, pp. 247–250.
37. Trond Ytterdal, Yuhua Cheng, Tor A. Fjeldly, *Device Modelling for Analog and RF CMOS Circuit Design*, John Wiley and Sons, USA, 2003.
38. Christian C. Enz, Eric A. Vittoz, *Charge-Based MOS Transistor Modeling: The EKV model for low-power and RF IC design*, John Wiley and Sons, USA, 2006.
39. Hyeokjae Lee, Kwun Soo Chun, Jeong Hyong Yi, Jong Ho Lee, Young June Park, and Hong Shick Min, "Harmonic Distortion due to narrow width effects in deep sub-micron SOI CMOS device for analog RF applications," *Proc. of EEE Int. Conf. on SOI*, Virginia, USA, 7–10 Oct. 2002, pp. 83–85.
40. Behzad Razavi, "A 300 GHz fundamental oscillator in 65-nm CMOS technology," *IEEE J. of Solid State Circuits*, vol. 46, no. 4, pp. 894–903, April 2011.

References

41. Ickjin Kwon and Kwyro Lee, "An accurate behavioral model for RF MOSFET linearity analysis," *IEEE Microwave and Wireless Components Letters*, vol. 17, no. 12, pp. 897–899, Dec. 2007.
42. Saptarsi Ghosh, Khomdram Singh, Sanjay Deb, and Subir Sarkar, "Two dimensional analytical modeling for SOI and SON MOSFET and their performance comparison," *J. of Nanoelectronics Physics*, vol. 3, no. 1, pp. 569–575, June, 2011.
43. S. Chouksey and J. G. Fossum, "DICE: a beneficial short-channel effect in nanoscale double-gate MOSFETs," *IEEE Trans. on Electron Devices*, vol. 55, no. 3, pp. 796–802, March 2008.
44. Viranjay M. Srivastava, K. S. Yadav, and G. Singh, "Analysis of double gate CMOS for double-pole four-throw RF switch design at 45-nm technology," *J. of Computational Electronics*, vol. 10, no. 1-2, pp. 229–240, June 2011.
45. Yang Tang, Li Zhang, and Yan Wang, "Accurate small signal modeling and extraction of silicon MOSFET for RF IC application," *Solid State Electronics*, vol. 54, no. 11, pp. 1312–1318, Nov. 2010.
46. Vaidyanathan Subramanian, "Multiple gate field effect transistor for future CMOS technologies," *IETE Technical Review*, vol. 27, no. 6, pp. 446–454, Dec. 2010.
47. S. M. Sze, *Semiconductor Devices: Physics and Technology*, 2nd Edition, Tata McGraw-Hill, India, 2004.
48. Yuhua Cheng, M. Deen, and Chih Chen, "MOSFET modeling for RF IC design," *IEEE Trans. on Electron Devices*, vol. 52, no. 7, pp. 1286–1303, July 2005.
49. John P. Uyemura, *Chip Design for Submicron VLSI: CMOS Layout and Simulation*, 1st Edition, Cengage Learning Publications, India, 2006.
50. W. Grabinski, *Transistor Level Modeling for Analog/RF IC Design*, 1st Edition, Springer Publications, Netherlands, 2006.
51. Didier Bouvet, Adrian Ionescu, Yusuf Leblebici, and Igor Stolitchnov, "Materials and devices for nanoelectronic systems beyond ultimately scaled CMOS," *Nanosystems Design and Technology*, vol. 1, pp. 23–44, 2009.
52. Mi Chang Chang, Chih Sheng Chang, Chung Lu, and Carlos H. Diaz, "Transistor and circuit design optimization for low power CMOS ," *IEEE Trans. on Electron Devices*, vol. 55, no. 1, pp. 84–95, Jan. 2008.
53. Robert Chau, Brian Doyle, Jack Kavalieros, and Kevin Zhang, "Integrated nanoelectronics for the future," *Nature Materials*, vol. 6, pp. 810–812, 2007.
54. H. C. Lo, C. T. Li, Y. T. Chen, C. T. Yang, W. C. Luo, W. Y. Lu, C. F. Cheng, T. L. Chen, C. H. Lien, H. T. Tsai, M. C. Chen, Samuel K. H. Fung, and C. C. Wu, "CMOS on dual SOI thickness for optimal performance," *Microelectronic Engineering*, vol. 87, no. 12, pp. 2531–2534, Dec. 2010.
55. A. Litwin, "Overlooked interfacial silicide-polysilicon gate resistance in MOS transistors," *IEEE Trans. on Electron Devices*, vol. 48, no. 9, pp. 2179–2181, Sept. 2001.
56. D. Lederer, C. Desrumeaux, F. Brunier, and J. P. Raskin, "High resistivity SOI substrates: how high should we go," *Proc. of IEEE Int. SOI Conf.*, California, USA, 29 Sept.-2 Oct. 2003, pp. 50–51.
57. Samar K. Saha, "Modeling process variability in scaled CMOS technology," *IEEE Design and Test of Computers*, vol. 27, no. 2, pp. 8–16, March–April 2010.
58. T. Manku, "Microwave CMOS device physics and design," *IEEE J. of Solid State Circuits*, vol. 34, no. 3, pp. 277–285, March 1999.
59. Viranjay M. Srivastava, K. S. Yadav, and G. Singh, "Application of VEE Pro software for measurement of MOS device parameter using C-V curve," *Int. J. of Computer Applications*, vol. 1, no. 7, pp. 43–46, March 2010.
60. Viranjay M. Srivastava, K. S. Yadav, and G. Singh, "Measurement of oxide thickness for MOS devices, using simulation of SUPREM simulator," *Int. J. of Computer Applications*, vol. 1, no. 6, pp. 66–70, March 2010.

61. Viranjay M. Srivastava, "Relevance of VEE programming for measurement of MOS device parameters," *Proc. of IEEE Int. Advance Computing Conf.*, India, 6-7 March 2009, pp. 205–209.
62. Y. Ye and Y. Cao, "Random variability modeling and its impact on scaled CMOS circuits," *J. of Computational Electronics*, vol. 9, no. 3, pp. 108–113, 2010.
63. R. H. Caverly, S. Smith, and J. Hu, "RF CMOS cells for wireless applications," *J. of Analog Integrated Circuits and Signal Processing*, vol. 25, no. 1, pp. 5–15, 2001.
64. Viranjay M. Srivastava, K. S. Yadav, and G. Singh, "Designing parameters for RF CMOS cells," *Int. J. of Circuits and Systems*, vol. 1, no. 2, pp. 49–53, Oct. 2010.
65. C. Laber, C. Rahim, S. Dreyer, G. Uehara, P. Kwok, and P. Gray, "Design considerations for a high performance 3 micron CMOS analog standard cell library," *IEEE J. of Solid State Circuits*, vol. 22, no. 2, pp. 181–189, Feb. 1987.
66. M. Smith, C. Portman, and C. Anagnostopoulos, "Cell libraries and assembly tools for analog/digital CMOS and BiCMOS ASIC design," *IEEE J. of Solid State Circuits*, vol. 24, no. 5, pp. 1419–1432, May 1989.
67. Y. Cheng and M. Matloubian, "Frequency dependent resistive and capacitive components in RF MOSFETs," *IEEE Electron Device Letters*, vol. 22, no. 7, pp. 333–335, July 2001.
68. K. J. Yang and C. Hu, "MOS capacitance measurements for high leakage thin dielectrics," *IEEE Trans. on Electron Devices*, vol. 46, no. 7, pp. 1500–1501, July 1997.
69. M. Smith, C. Portman, C. Anagnostopoulos, P. Valdenaire, and H. Ching, "Analog CMOS integrated circuit design: research and undergraduate teaching," *IEEE Trans. on Education*, vol. 32, no. 3, pp. 210–217, Aug. 1989.
70. T. Manku, "A small-signal MOSFET model for radio frequency IC applications," *IEEE Trans. on Computer Aided Design of Integrated Circuits and Systems*, vol. 16, no. 5, pp. 437–447, May 1997.
71. K. Bernstein, "High-performance CMOS variability in the 65 nm regime and beyond," *IBM J. Res. Dev.*, vol. 50, no. 4, pp. 433–449, 2006.
72. N. Talwalkar, C. Patrick Yue, and S. Simon Wong, "Integrated CMOS transmit-receive switch using LC tuned substrate bias for 2.4 GHz and 5.2 GHz applications," *IEEE J. of Solid State Circuits*, vol. 39, no. 6, pp. 863–870, June 1989.
73. A product manual, *RF MOSFET*, Infineon Technologies, USA, 2010.
74. D. Su, M. Zargari, P. Yue, D. Weber, B. Kaczynski, and B. Wooley, "A 5 GHz CMOS transceiver for IEEE 802.11a wireless LAN systems," *IEEE J. of Solid State Circuits*, vol. 37, no. 12, pp. 1688–1694, Dec. 2002.
75. M. Uzunkol and G. M. Rebeiz, "A low loss 50–70 GHz SPDT switch in 90 nm CMOS," *IEEE J. of Solid State Circuits*, vol. 45, no. 10, pp. 2003–2007, Oct. 2010.
76. Kenneth L. Kaiser, *Electrostatic Discharge*, 1st Edition, Taylor and Francis, USA, 2006.
77. G. A. Brown, P. M. Zeitzoff, G. Bersuker, and H. R. Huff, "Mobility evaluation in transistors with charge-trapping gate dielectrics," Appl. Phys. Lett., vol. 87, no. 4, pp. 1–3, 2005.
78. C. Chiu, "A sub 400 ^{0}C Germanium MOSFET technology with high-k dielectric and metal gate", *Proc. of Int. Electron Device Meeting*, San Francisco, USA, 8–11 Dec. 2002, pp. 437–440.
79. H. Shang, "High mobility p-channel Germanium MOSFETs with a thin Ge-Oxynitride gate dielectric," *Proc. of Int. Electron Device Meeting*, San Francisco, USA, 8–11 Dec. 2002, pp. 1–4.
80. R. Chau, "Advanced depleted substrate transistor: single-gate, double-gate, and tri-gate," *Solid State Device Meeting*, 2002, pp. 68–69.
81. Fu Liang Yang, "25 nm CMOS Omega FETs," *Proc. of Int. Electron Device Meeting*, San Francisco, USA, 8–11 Dec. 2002, p. 255.
82. J. Colinge, "Silicon-on-insulator Gate-all-around Device," *Proc. of Int. Electron Device Meeting*, San Francisco, USA, 1–3 Dec. 1990, p. 595.

References

83. J. Kedzierski, "Metal-gate FinFET and fully-depleted SOI devices using total gate silicidation," *Proc. of Int. Electron Device Meeting*, San Francisco, USA, 1–3 Dec. 1990, p. 247.
84. T. Park, "Fabrication of Body-Tied FinFETS (Omega MOSFETS) Using Bulk Si Wafers," *Symp. on VLSI Technology*, Kyoto, Japan, 10–12 June 2003, pp. 135–136.
85. S. Monfray et. al., "50 nm – Gate all around (GAA) – Silicon on nothing (SON) – devices: A simple way to co-integration of GAA transistors with bulk MOSFET process," Symposium on *VLSI technology Digest of Technical Paper*, Honolulu, USA, 11–13 June 2002, pp. 108–109.
86. Sung Young Lee, Sung Min Kim, Eun Jung Yoon, Chang Woo Oh, Ilsub Chung, Donggun Park, and Kinam Kim, "A novel multibridge-channel MOSFET (MBCFET): fabrication technologies and characteristics," *IEEE Trans. on Nanotechnology*, vol. 2, no. 4, pp. 253–257, April 2003.
87. Xuejie Shi and Man Wong, "Effects of substrate doping on the linearly extrapolated threshold voltage of symmetrical DG MOS devices," *IEEE Trans. on Electron Devices*, vol. 52, no. 7, pp. 1616–1621, July 2005.
88. K. W. Guarini, "Triple-self-aligned, planar double-gate MOSFETs: devices and circuits," *Proc. of Int. Electron Device Meeting*, Washington DC, USA, 2–5 Dec. 2001, pp. 19.2.1–19.2.4.
89. B. Goebel, "Fully depleted surrounding gate transistor (SGT) for 70 nm DRAM and beyond," *Proc. of Int. Electron Device Meeting*, San Francisco, USA, 8–11 Dec. 2002, pp. 275–278.
90. Meishoku Masahara, "15-nm thick Si channel wall vertical double-gate MOSFET," *Proc. of Int. Electron Device Meeting*, San Francisco, USA, 8–11 December 2002, pp. 949–951.
91. Serge Gidon, "Double-gate MOSFET modeling," *Proc. of the COMSOL Multiphysics User's Conf.*, Paris, 2005, pp. 1–4.
92. Viranjay M. Srivastava, K. S. Yadav, and G. Singh, "Drain current and noise model of cylindrical surrounding double-gate MOSFET for RF switch," *Procedia Engineering*, vol. 38, pp. 517–521, April 2012.
93. A. Nitayami, H. Takato, N. Okabe, K. Sunouchi, K. Hiea, and F. Horiguchi, "Multipillar surrounding gate transistor (M-SGT) for compact and high-speed circuits," *IEEE Trans. on Electron Devices*, vol. 38, pp. 579–583, 1991.
94. S. Watanabe, K. Tsuchida, D. Takashima, Y. Oowaki, A. Nitayama, and K. Hieda, "A novel circuit technology with surrounding gate transistors (SGT's) for ultra high density DRAM's," *IEEE J. Solid State Circuits*, vol. 30, no. 9, pp. 960–971, Sept. 1995.
95. F. Djeffal, Z. Ghoggali, Z. Dibi, and N. Lakhdar, "Analytical analysis of nanoscale multiple gate MOSFETs including effects of hot carrier induced interface charges," *Microelectronics Reliability*, vol. 49, no. 4, pp. 377–381, April 2009.
96. Viranjay M. Srivastava, K. S. Yadav, and G. Singh, "Explicit model of cylindrical surrounding double-gate MOSFETs," *WSEAS Trans. on Circuits and Systems*, vol. 12, no. 3, pp. 81–90, March 2013.
97. Cong Li, Yiqi Zhuang, and Ru Han, "Cylindrical surrounding-gate MOSFETs with electrically induced source/drain extension," *Microelectronics Journal*, vol. 42, no. 2, pp. 341–346, Feb. 2011.
98. Te Kuang Chiang, "Concise analytical threshold voltage model for cylindrical fully depleted surrounding-gate MOSFET," *Jpn. J. Appl. Phys.*, vol. 44, no. 5, pp. 2948–2952, 2005.
99. S. Cristoloveanu and S. S. Li, *Electrical Characterization of SOI Materials and Devices*, Kluwer Publications, Massachusetts, USA, 1995.
100. H. S. Baik and S. J. Pennycook, "Interface structure and non-stoichiometry in HfO_2 dielectrics," *IEEE Applied Physics Letter*, vol. 85, pp. 672–674, 2009.
101. J. C. Lee, "Single-layer thin HfO_2 gate dielectric with n^+ poly-Silicon," *Proc. of IEEE Symposium on VLSI Technology*, Honolulu, USA, 13–15 June 2000, pp. 44–45.
102. Thomas Sokollik, "Plasma Physics," *Investigations of Field Dynamics in Laser Plasmas with Proton Imaging*, vol. 1, pp. 17–24, 2011.

103. Ta Chang Tien, Li Chuan Lin, Lurng Shehng Lee, Chi Jen Hwang, Siddheswar Maikap, and Yuri M. Shulga, "Analysis of weakly bonded oxygen in $HfO_2/SiO_2/Si$ stacks by using HRBS and ARXPS," *J. of Material Science: Material Electronics*, vol. 21, no. 5, pp. 475–480, 2010.
104. A. P. Huang, Z. C. Yang, and Paul K. Chu, "Hafnium based high-k gate dielectrics," *Advances in Solid State Circuits Technologies*, pp. 333–350, April 2010.
105. C. Wang and J. Hwu, "Characterization of stacked Hafnium-oxide (HfO_2) / Silicon-dioxide (SiO_2) metal-oxide-semiconductor tunneling temperature sensors" *J. of Electrochemical Society*, vol. 157, no. 10, pp. 324–328, 2010.
106. M. Fadel, and O. Azim, "A study of some optical properties of hafnium dioxide (HfO_2) thin films and their applications," *J. of Applied Physics Materials Science and Processing*, vol. 66, no. 3, pp. 335–343, 1997.
107. P. Dutkiewicz, M. Kieczewski, K. Kozowski, and D. Pazderski, "Vision localization system for mobile robot with velocities and acceleration estimator," *Bulletin of the Polish Academy of Sciences and Technical Sciences*, vol. 58, no. 1, pp. 29–41, Dec. 2010.
108. W. Jendernalik, J. Jakusz, G. Blakiewicz, R. Piotrowski, and S. Szczepanski, "CMOS realisation of analogue processor for early vision processing," *Bulletin of the Polish Academy of Sciences and Technical Sciences*, vol. 59, no. 2, pp. 141–147, Aug. 2011.
109. D. A. Martin, H. S. Lee, and I. Masaki, "A mixed signal array processor with early vision applications," *IEEE J. Solid State Circuits*, vol. 33, no. 3, pp. 497–502, March 1998.
110. P. Dudek, A. Lopich, and V. Gruev, "A pixel parallel cellular processor array in a stacked three layer 3D silicon-on-insulator technology," *Proc. of Eur. Conf. on Circuit Theory and Design*, Antalya, Turkey, 23–27 Aug. 2009, pp. 193–196.
111. E. Culurciello and P. Weerakoon, "Three dimensional photo detectors in 3D silicon-on-insulator technology," *IEEE Electron Device Letters*, vol. 28, pp. 117–119, 2007.
112. G. Blakiewicz, "Analog multiplier for a low-power integrated image sensor," *16th Int. Conf. on Mixed Design of Integrated Circuits and Systems*, 25–27 June 2009, Poland, pp. 226–229.
113. A. G. Andreou, R. C. Meitzler, K. Strohbehn, and K. A. Boahen, "Analog VLSI neuromorphic image acquisition and pre-processing systems," *Neural Networks*, vol. 8, no. 7–8, pp. 1323–1347, 1995.
114. M. Furumiya, "High sensitivity and no-crosstalk pixel technology for embedded CMOS image sensor," *IEEE Trans. Electron Devices*, vol. 48, no. 10, pp. 2221–2227, Oct. 2001.
115. Maria Petrou, and Panagiota Bosdogianni, *Image processing: The fundamental*, John Wiley and Sons, USA, 2000.
116. Viranjay M. Srivastava, K. S. Yadav, and G. Singh, "DP4T RF CMOS switch: A better option to replace SPDT switch and DPDT switch," *Recent Patents on Electrical and Electronic Engineering*, vol. 5, no. 3, pp. 244–248, Oct. 2012.

Chapter 2
Design of Double-Pole Four-Throw RF Switch

2.1 Introduction

The industrial, scientific, and medical (ISM) radio bands were originally reserved for the use of radio-frequency (RF) energy for industrial, scientific, and medical purposes such as radio-frequency process heating, microwave ovens, and medical diathermy machines. The powerful emissions of these devices can create electromagnetic interference and disrupt radio communication using the same frequency, so these devices were limited to certain bands of frequencies. In general, communication equipment operating in these bands must accept any interference generated by ISM equipment [1, 2]. Nowadays CMOS wideband switches are designed primarily to meet the requirements of devices transmitting at ISM band frequencies (900 MHz and above). The low insertion loss, high isolation between ports, low distortion, and low current consumption of these devices make them an excellent solution for several high-frequency applications [3].

This chapter begins with an introduction to the various families of switch topologies used in earlier researches [4–6]. Specifications for an integrated CMOS switch are then developed, followed by a description and working of the topology used in this book. Based on earlier reported works [7] and datasheets of existing parts [8], a set of specifications are analyzed for an integrated transceiver switch for advanced wireless local area network (LAN) systems [9, 10]. The asymmetry in the received and transmitted power levels can be effectively used to arrive at a set of specifications which can be met by designs in a CMOS technology.

2.2 Comparison of Various Switches

There are varieties of switches available in the market; however, the most important and basic types of switches are discussed as follows.

2.2.1 PIN Diode Switch

A PIN diode is a semiconductor diode in which a high resistivity intrinsic (I) region is sandwiched between p-type and n-type region and named as P-I-N. Without bias, the diode behaves like a capacitance and with biasing the diode behaves like an inductor. It has a very significant linearity, so it can be used for high-frequency (HF) applications and very high power applications; however, as the higher DC power applied, the lower insertion loss occurs. PIN diode is a current-controlled resistance that operates as a variable resistor at RF and microwave frequency regime of the electromagnetic spectrum. The resistance value of the PIN diode is determined only by the forward-biased DC current. An additional feature of the PIN diode is its ability to control large RF signals while using much smaller levels of DC excitation. In the ON-state, the diode is biased using a large current of about 10 mA which ensures that the AC resistance is low [8]. In the OFF-state, the PIN structure has a low junction capacitance which ensures the large isolation. PIN diodes can be fabricated in silicon and gallium arsenide and it shows very significant insertion loss (<1 dB) and power handling (>5 W) up to very high frequencies; their static power consumption due to the bias current remains a severe limitation [9]. Since a large bias current is typically required for switch operation, it must be supplied through a choke. Due to limitations of static power consumption, diode switches are being gradually replaced by GaAs MESFET, which offers only slightly worse performance for significantly lower static power consumption.

2.2.2 GaAs FET Switch

A GaAs-integrated circuit switch is an integrated circuit using field-effect transistors to achieve switching between multiple paths [10]. It acts as a voltage-controlled resistor, used for the broadband usually from 0.5 to 4 GHz. This switch specially tuned for application at 5–6 GHz, with low insertion loss of 0.25 dB with good isolation performances and low power consumption and better switching speed. However, it can work up to maximum control voltage of 6 V. To avoid the above disadvantages of the PIN diode switch and GaAs switch, we proposed the CMOS switch. A CMOS switch is an integrated circuit using MOSFET to achieve switching between multiple paths [11]. It has low dependency of the P_1 dB versus control voltage with better switching speed. Other parameters for the CMOS switch are discussed in this chapter and following chapters.

2.2.3 MESFET Switch

The Metal Semiconductor Field Effect Transistor (MESFETs) are majority carrier devices and applicable for high-speed operation. It can be implemented using silicon, gallium arsenide, and indium phosphide [12]; however, silicon-based

MESFETs are incapable of handling large powers and typically slower than those implemented using the other materials. For high-power (>1 W) and high-frequency (>1 GHz) applications, the MESFETs are implemented using GaAs. This is due to large band gap of GaAs, and hence a large breakdown voltage that allows high voltage operation with no reliability concerns can be achieved [13]. For the design of a GaAs MESFET, the main trade-off is between its ON-state resistance and OFF-state capacitance. In order to achieve a low insertion loss, a large device with a low ON-resistance can be used. This degrades the isolation performance since the OFF-state capacitance will be large. An important limitation is that they cannot be integrated with silicon-based transceivers. Another limitation of GaAs MESFET switches is their power handling capability, as compared to PIN diodes [8, 9]; GaAs MOSFET switches do not consume static power, which makes them attractive for low-power wireless communication devices.

2.2.4 MOSFET Switch

The Metal-Oxide-Semiconductor Field-Effect-Transistor (MOSFET) is one of the simplest switch options. It is available in a Complementary Metal-Oxide-Semiconductor (CMOS) process, and its performance improves every decade. Only silicon-based MOSFETs are suitable for the switch due to the absence of a gate insulator for other materials [14]. The ON-resistance of silicon MOSFET is significantly inferior to a GaAs MESFET due to poor electron and hole mobility in channel at low electric fields. Recent technology offers very small channel length MOSFETs with a better $R_{ON} \times C_{OFF}$ product [15]. The thin gate dielectric and small channel length permit a low voltage operation. The channel resistances and the substrate resistances are the main sources of power loss in the MOSFET. The substrate resistance may be reduced by grounding the substrate as close to the device as possible. The low quality factor of the source and drain parasitic junction capacitors can also lead to significant losses, especially as the frequency of operation increases. The linearity of the MOSFET switch is limited for large signal swings due to conductivity modulation caused by a changing gate–source (V_{gs}) and drain–source (V_{ds}) voltage for a large signal input.

2.2.5 MEMS Switch

The Micro-Electro-Mechanical (MEMS) switches are micro-machined devices which use a mechanical and physical movement to achieve ON (short) or OFF (open) circuit in the transmission line [16]. The mechanical and physical movement of MEMS switch controls the impedance of a transmission line. Generally RF MEMS switches are designed to operate in millimeter wave or microwave frequencies (0.1–100 GHZ) regime of the spectrum. The RF MEMS switches have an advantage over the traditional RF switches with their broadband operation, high

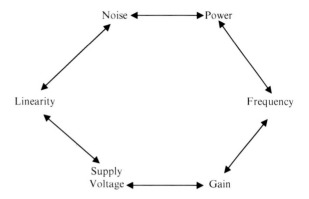

Fig. 2.1 Radio-frequency design hexagon

isolation, low insertion loss, low power consumption, simple biasing networks, and low intermodulation products [17]. However they have several disadvantages like slow switching speed in the orders of microsecond, high actuation voltage requirement, and hot switching effects in high-power applications.

RF circuits suffer from trade-offs among various parameters as shown in Fig. 2.1 an RF design hexagon, where six important circuit parameters are shown to trade-off each other. It is interesting to point out that in some cases (e.g., power amplifiers) if the supply voltage is reduced, the power dissipation may increase [18]. For this reason, supply scaling in RF circuits lags behind that in digital circuits. The RF design hexagon also indicates that simple figures of merit such as the transit frequency, unity power gain frequency, and gate delay cannot be easily used to predict RF performance because they do not reflect many of the trade-offs [19].

2.3 RF Transceiver Systems

A superheterodyne RF transceiver architecture contains a few blocks which are implemented off-chip, as shown in Fig. 2.2a. This includes the antenna, preselection band-pass filter, intermediate frequency filter, and transceiver switch. While efforts are being made to integrate these blocks on a single chip using standard CMOS technologies [19–21], the quality factor of on-chip inductors and the substrate parasitics of MOSFETs are also very important limiting factors.

Earlier, the reported literatures [3, 22] suggest that receivers are being integrated from the low-noise amplifier onwards while the transmitters are integrated up to the power amplifier. An integrated transceiver switch which includes matching networks for the low-noise amplifier and power amplifier will push the integration boundary further towards the antenna as shown in Fig. 2.2a. However, such an improvement also decreases board component count and hence total cost. Therefore, the transceiver switch is a desirable, as well as a suitable candidate for evaluating the impact of the inductive substrate bias technique.

The purpose of a transceiver switch is to alternately couple the antenna to either the transmitter or the receiver and to protect the receiver while transmitting high power.

2.3 RF Transceiver Systems

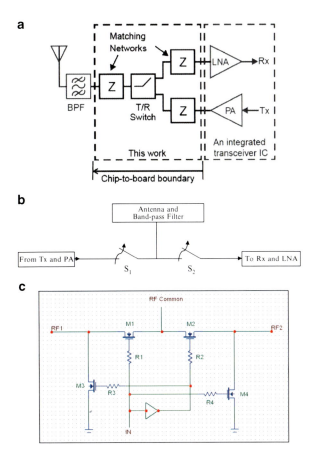

Fig. 2.2 A radio front-end block diagram with (**a**) the integration of transceiver switch and matching networks, (**b**) simplified schematic of a transceiver switch, and (**c**) typical transistor based transceiver switch [22]

The effect of integrating the transceiver switch is to push the chip-to-board boundary closer to the antenna. A simple schematic of a transceiver switch is shown in Fig. 2.2b. The switch operates in either the transmit (T_x) mode, in which power is transmitted from the power amplifier (PA) to the antenna, or in the receive (R_x) mode, when power is delivered from the antenna to the LNA [23–25].

These two switches, S_1 and S_2, are operated by using opposite phases of control signal (V_{ctrl}), thereby ensuring that the antenna is connected to either the LNA or the PA, but not both. Ideally, for all input power levels, the S_1 switch and S_2 switch would be perfect "short-circuited" with zero impedance when closed and perfect "open-circuited" with infinite impedance when the contacts are separate or open. To improve their bandwidth, wideband switches use only n-channel MOSFETs in the signal path. An n-type MOSFET switch has a typical −3 dB bandwidth of 400 MHz, almost twice the bandwidth performance of a standard switch with n-type MOS transistor and p-type MOS transistor in parallel. This is a result of the smaller switch size and greatly reduced parasitic capacitance due to removal of the p-channel MOSFET. The n-channel MOSFETs act essentially as voltage-controlled resistors. The switches operate as follows:

$$V_{gs} > V_{th} \rightarrow \text{Switch ON}$$

$$V_{gs} < V_{th} \rightarrow \text{Switch OFF}$$

where V_{gs} is the gate-to-source voltage and V_{th} is defined as the threshold voltage above which a conducting channel is formed between the source and drain terminals. As the signal frequency increases greater than several hundred MHz, parasitic capacitances tend to dominate. Therefore, high isolation in the switch OFF-state and low insertion loss in the switch ON-state can be achieved. For the wideband applications, it is a quite challenging task for the switch designers. The channel resistance of a switch should be limited to less than about 6 Ω to achieve a low-frequency insertion loss of less than 0.5 dB on a line with 50 Ω matched impedances at the source and load [24, 26, 27].

2.4 RF Transceiver Switch

For transmitting process, as the voltage V_1 goes high (+5 V or high level) and voltage V_2 goes low (−5 V or low level), then these condition turns transistor M_1 and transistor M_4 in ON-state and transistor M_2 and transistor M_3 in OFF-state. For receiving process, as the voltage V_1 goes low (−5 V or low level) and voltage V_2 goes high (+5 V or high level), then these condition turns transistor M_1 and transistor M_4 in OFF-state and transistor M_2 and transistor M_3 in ON-state. Transistor M_3 and transistor M_4 shunt the signal in receive and transmit mode, respectively, and thus increase the isolation. Capacitance C_1 and C_2 allow DC biasing of the transmitting and receiving nodes. The purpose of resistance R_1, R_2, R_3, and R_4 is to improve DC bias isolation and has a value of about 10 kW. This circuit has very good isolation in OFF-state but suffer from high loss in ON-state because of the shunt transistors. It also has nonlinear properties when the power of the signal increases.

The FETs have an interlocking finger layout that reduces the parasitic capacitance between the input (RF$_X$) and the output (RF$_C$), thereby increasing isolation at high frequencies and enhancing cross talk rejection. For example, when M_1 is ON to form the conducting path for RF$_1$, M_2 is OFF and M_4 is ON, shunting the parasitics at RF$_2$ to ground, as shown in Fig. 2.2c. In the normal operation, the switches can handle a 7 dBm (5 mW) input signal. For a 50 Ω load, this corresponds to a 0.5 V rms signal or 1.4 V peak to peak for sine waves. The power handling capability is reduced at lower frequencies for the following two reasons:

a. Since the n-type MOSFET structure consists of two regions of n-type material in a p-type substrate. Parasitic diodes are thus formed between the n and p regions. When an AC signal, biased at nearly zero volts DC, is applied to the source of the transistor, and V_{gs} is large enough to turn the transistor ON ($V_{gs} > V_{th}$), the parasitic diodes can be forward biased for some portion of the negative half-cycle of the input waveform. This happens if the input sine wave goes below approximately 0.6 V, and the diode begins to turn ON, thereby causing the input signal to be clipped (compressed).

b. The less power handling capability at lower frequencies is the partial turn-ON of the shunt n-type MOSFET device when it is supposed to be OFF. This is very similar to the mechanism described above where there was partial turn-ON of parasitic diode. In this case, the n-type MOSFET transistor is in the OFF-state, by the way of $V_{gs} < V_{th}$. With an AC signal on the source of the shunt device, there will be a time in the negative half-cycle of the waveform, where $V_{gs} > V_{th}$, thereby partially turning ON the shunt device. This will compress the input waveform by shunting some of its energy to ground [31, 32].

In the communication systems, to transmit or receive the information through multiple antennas such as multiple-input and multiple-output (MIMO) systems, it is essential to design a novel RF switch that is capable of operating with multiple antennas and frequencies as well as minimizing signal distortion and power consumption [28–30]. The Si-CMOS for this application allows the higher levels of integration and lower cost as well as improvement in the efficiency. In a switch, the modulated signal is simply transmitted through the switch and makes its way to the antenna for releasing into the space. On the other hand, the modulated signal is received by the antenna and makes its way through the switching path to the receiver. Currently, the CMOS technology is almost capable to take overall functionalities of radio-frequency circuitry.

2.5 Design of CMOS Inverter for RF Switch

The CMOS-based technology is the main contributor to reducing significantly the switching time in digital circuits and high-speed performance of the analog electronics. The low-power dissipation and integrability on a single chip made this a demanding technology. Compared to the single-transistor gate logic which consists of n-type MOS transistor with a resistor pull-up and the BJT-based TTL, the CMOS established a new paradigm for low power consumption and high operating speed. The new advances in MOSFET technologies such as double-gate, gate-all-around, and FinFET are the most promising configurations. In this book we have discussed about the double-gate MOSFET and a new proposed design of cylindrical surrounding double-gate MOSFET.

The reason for the success of Silicon MOSFET technology is the development of CMOS logic, because this technology provides both n-channel and p-channel MOSFET. The basic MOSFET has four terminals:

a. Source
b. Drain

Fig. 2.3 Schematic of the CMOS (**a**) internal structure and (**b**) inverter circuit [31]

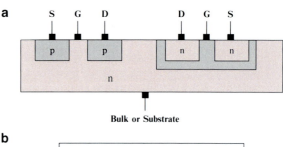

c. Gate
d. Bulk

Figure 2.3 shows the cross-sectional view of a typical CMOS process. In the n-type MOS transistor, if the voltage on the gate increases, the majority carriers (holes) will be pushed away towards substrate and have been depleted as gate voltage continues to increase. Eventually, generations of carriers will exceed the recombination. The generated hole–electron pairs are separated by the field, the holes being swept into the bulk and the electrons moving to the oxide–silicon interface, where they are held, because of the energy barrier between the conduction band in the silicon and that in the oxide. When a drain-to-source voltage is imposed, the current will flow in the channel. As the drain-to-source voltage (V_{ds}) increases, the current also increases. However, beyond a certain drain-to-source voltage, the current will saturate.

The p-channel MOSFET relies on n-type substrate. As commonly p-type wafers are used for processing, an additional n-type well implant is necessary. In this well, which is a deep region of n-type doping, the p-channel MOSFET is placed. As the p-substrate and the n-well junction are reverse biased, therefore no significant current flows between these regions and the two transistors are isolated as shown in Fig. 2.3a. The output current of the p-channel MOSFET is typically much lower than the current of an n-channel MOSFET with similar dimensions and doping. This is due to the lower carrier mobility of holes compared to electrons. As the characteristics of the complementary transistors should be as equal as possible, the width of the p-channel MOSFET is typically made larger to compensate the difference. We have taken the geometry factor (ratio of p-type MOSFET to n-type MOSFET) equal to 3.5 to obtain the equal drain currents for equal gate biases of the device [31].

In the stationary case as in Fig. 2.3b, the circuit does not consume any power when assuming perfect devices without leakage current. It can be seen that the gates are at the same bias V_{in} which means that they are always in a complementary state. When V_{in} is high, $V_{in} \approx V_{dd}$, the voltage between gate and substrate of the n-type MOS transistor is also approximately V_{dd} and the transistor is in ON-state. The gate–substrate bias at the p-type MOS transistor on the other side is nearly zero and the transistor is turned OFF. The output voltage V_{out} is pulled to ground, which is the low state. When V_{in} is low $V_{in} \approx 0$, the complementary situation occurs and the p-MOSFET is turned ON, while the n-type MOSFET is turned OFF. The output voltage is, therefore, pulled to V_{dd} which is the high state. It is important to note that in both states, high and low, no static current flows through the inverter. This is valid when assuming ideal devices with zero OFF-current and leakage current [27, 32]. Considering the negative bias temperature instability, the worst stress conditions are imposed on the p-channel MOSFET at $V_{in} = V_{low}$. At this bias condition the p-type MOSFET is turned ON, with approximately the same potential at the source and the drain $V_{gs} = V_{gd} = V_{dd}$ and negative gate-to-substrate voltage $V_{gsub} = -V_{dd}$.

2.6 Configuration of Switches

The RF switch is one of the key functional building blocks. There is a new trend to employ the CMOS technique to control the conducting path between transmitter/receiver circuit and antenna. By this way, it will be helpful to the integration of RF switch with CMOS transceiver circuit, reducing the overall cost of the chipset [33, 34]. There are several types of RF switches, which have been classified as follows.

2.6.1 Single-Pole Single-Throw Switch

The single-pole single-throw (SPST) switch plays an important role in communication and radar systems. It can be used as an individual component or an integral element in subsystems or systems, for example, in RF pulse transmitters [2]. For high data rate short-range communication and some high-resolution radar systems, the SPST switches used for the transmitters pulse formation with fast switching time and high isolation including low insertion loss. Among these characteristics,

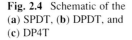
Fig. 2.4 Schematic of the (a) SPDT, (b) DPDT, and (c) DP4T

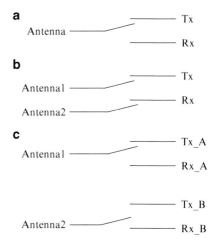

the high isolation is especially a crucial in order to reduce or prevent RF leakage [24, 26, 27]. However, the undesired RF leakage not only causes external effects, such as harm to other coexisting systems, but also internal effects such as reduced dynamic range.

2.6.2 Single-Pole Double-Throw Switch

The single-pole double-throw (SPDT) switch is the fundamental switch that links between one antenna and the transmitter/receiver as shown in Fig. 2.4a. In time division duplexing (TDD) communication systems, transmitter/receiver switch plays an important role to change the RF signal flow to the transmitter or receiver. Further, to increase the integration level, the SPDT switch has to be integrated in the transceiver. The advantages of silicon CMOS technology for RF and microwave control functions over GaAs are its low cost and the integration potential with RF and silicon MOS-based mixed-signal circuitry [27].

Due to the limitations of the CMOS process and circuit topology, the frequencies of most reported CMOS RF switches are lower than 5.8 GHz [28]. The conventional topology for CMOS switch is series shunt, which is only suitable for narrow band design. It demonstrates the power performance of a CMOS switch, but the LC-tuned substrate bias network limits the frequency response. For broadband frequency response, a switch using 0.13-μm CMOS process based on traveling wave concept is reported [35]. The switch consist of the OFF-state shunt transistors and series micro-strip lines to form an artificial transmission line with 50 Ω characteristic impedance and achieve wide bandwidth.

Recently, Quemerais et al. [36] have proposed a fully integrated SPDT transceiver switch, which has been implemented on a standard 45-nm CMOS process.

2.6 Configuration of Switches

This circuit is dedicated to fully integrated CMOS front-end modules operating at GHz range. Dinc et al. [37] have proposed an SPDT transceiver switch for X-band on-chip radar applications. These methodologies include the optimization of transistor widths for significantly lower insertion loss, while preserving high isolation and using a parallel resonance technique to improve the isolation. Also, the techniques such as applying DC bias to the source and the drain, using on-chip impedance transformation networks and the body floating are used to improve the power handling capability of the switch. Lei et al. [38] have designed an SPDT switch in a partially depleted (PD) SOI CMOS process for 2.4 GHz wireless communication applications. However, based on the advantage of PD SOI device structure, the presented switch demonstrates the high performance on insertion loss and isolation. Maisurah et al. [39] have designed SPDT transceiver switch for 900 MHz frequency applications with a 0.18-µm CMOS process. Mekanand et al. [29] have proposed a transceiver CMOS switch for 2.4 GHz with low insertion loss and excellent control voltage [40, 41]. These simulation results of CMOS switch design demonstrate an insertion loss of 1.102 dB for receiving mode and 1.085 dB for transmitting mode. However, both the modes can operate using a control voltage of only 1.2 V.

2.6.3 Double-Pole Double-Throw Switch

The SPDT is the fundamental switch that links between antenna and the analog front-end section but due to the single operating frequency. This type of switch has a limited data transfer rate. Therefore, a double-pole double-throw (DPDT) switch is developed to solve the aforementioned problem as shown in Fig. 2.4b. The preferred DPDT switch is the back-to-back structure, which combines the common ports of conventional SPDT switches because it has only one device between four ports. Hence this reduces the insertion loss by half as compared to the SPDT that has two devices in each signal path. In the wireless communication systems, the demand for $n \times m$ switch matrices is increasing significantly for the antenna diversity. The general switching matrix has the disadvantage of the need for large number of switching devices and complicated control logic. These DPDT switches has dual antenna and dual ports, one for transmitting and another for receiving, which is not sufficient for MIMO systems [42]. Hence, we have designed a double-pole four-throw (DP4T) switch to enhance the switch performance for MIMO communication applications.

2.6.4 Double-Pole Four-Throw Switch

This double-pole four-throw (DP4T) switch can send or receive two parallel data streams simultaneously as shown in Fig. 2.4c [5]. The switch with the CMOS inverter technology is shown in Fig. 2.5 to design a novel DP4T switch and then

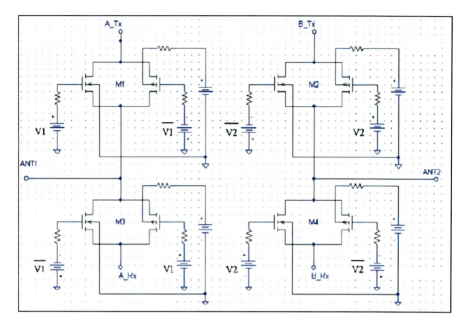

Fig. 2.5 DP4T CMOS transceivers switch with single-gate transistor

design this DP4T switch using the double-gate (DG) MOSFET in the following chapters. Its performances, such as insertion loss, control power, and signal distortion, are compared with a traditional n-type MOSFET DP4T switch in terms of single-gate and double-gate [27, 40, 41].

2.7 Design of DP4T RF Switch Based on Single-Gate MOSFET

Already existing designs of the DP4T switches use the single-gate n-type MOSFET [29] and require a high control voltage from 3.0 to 5.0 V to operate, and a large internal/contact resistance and capacitances are produced for this design at the receivers, transmitters, and antennas for detecting the data signal as shown in Fig. 2.5. This high value of control voltage is not suitable for modern low power portable devices which require smaller power consumption of about 0.5–2.1 V. Therefore, we have proposed a novel DP4T switch using DG CMOS technology for the purpose of RF application [42, 43].

The objective of the proposed design of a switch is to operate at 2.4 and 5.0 GHz for MIMO systems. This switch mitigates the attenuation of the passing signals and exhibits high isolation to avoid interruption of the simultaneously received signals [29]. Since in Fig. 2.5, four transistors are used for first antenna, the working process

2.7 Design of DP4T RF Switch Based on Single-Gate MOSFET

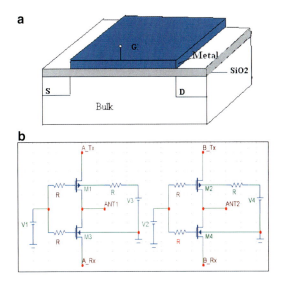

Fig. 2.6 Schematic of the (**a**) basic SG MOSFET and (**b**) DP4T SG RF CMOS switch

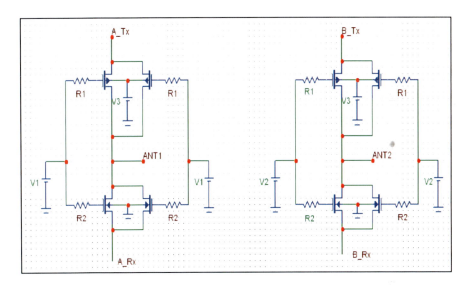

Fig. 2.7 Proposed DP4T switch with two transistors

at a time any one of transistor M_1 or transistor M_3 will operate and in the same fashion any one of transistor M_2 or transistor M_4 will operate. Similar working function is observed in the proposed DP4T CMOS switch as in Figs. 2.6b and 2.7.

A double-pole four-throw double-gate radio-frequency complementary metal-oxide-semiconductor (DP4T DG RF CMOS) switch has the properties as fixed-tuned matching networks, low quality factor, high output power, power gain, or power amplification, noise figure or amount of noise added during normal operation, and high-power dissipation (total power consumption). Some bipolar RF CMOS transistors are suitable for automotive, commercial, or general industrial applications [43, 44]. To design the DP4T switch based on the DG MOSFET, first we design the DP4T switch in the present chapter with the two parallel transistor technique and then designed the DG MOSFET as in Chap. 3 and at last combine the both technology to design the proposed model of DP4T DG RF CMOS switch in Chap. 4.

The choice of RF CMOS switch requires an analysis of the performance parameters such as maximum drain saturation current, operating frequency, cut-off frequency, threshold voltage of n-type MOSFET and p-type MOSFET, control voltage, output power, and forward transconductance [45]. It also controls the increase or decrease of channel lengths for the devices which operates in depletion region. The DP4T switch is a fundamental switch for the application of multiple-input, multiple-output data transfer. So, the parallel data streams can be transmitted or received simultaneously using the multiple antennas.

2.8 Operational Characteristics of DP4T CMOS Switch

Figure 2.5 shows the existing DP4T CMOS transceiver switch with single-gate (SG) transistor [29]. Here two antennas and four ports are taken into account. Figure 2.6a shows the SG MOSFET structure. In Fig. 2.6b, the transmitted signal from power amplifier is sent to transmitter "A" named as "A_T_x"port and travel to the ANT_1 node while the received signal will travel from the ANT_2 node to the receiver "B" named as "B_R_x" port and pass on to the low-noise amplifier (LNA) or any other application as required for transceiver systems. The proposed switch contains CMOS in its architecture and needs only two control lines (V_1, V_2) of 1.2 or 2.1 V to control the signal congestion between two antennas and four ports; therefore, it improves the port isolation performance two times as compared to the DPDT switch and reducing signal distortion. Since connecting an n-type MOSFET in series with a p-type MOSFET allows signals to pass in either direction as shown in the Fig. 2.7. Whether the n-type or p-type device carries more signals current depends on the ratio of input to output voltage because the switch has no preferred direction for the current flow, so it has no preferred input or output terminals.

When the low voltage (approximately zero volts) is applied at the input of Fig. 2.3b, the top transistor (p-type) is conducting (switch closed) while the bottom transistor behaves like an open circuit. Therefore, the supply voltage (5 V) appears at the output. Conversely, when a high voltage (5 V) is applied at the input, the bottom transistor (n-type) is conducting (switch closed) while the top transistor behaves like an open circuit. Hence, the output voltage is low (0 V). It is important to note that always one of the transistors will be an open circuit and no current flows from the supply voltage to the ground [46, 47]. The voltage transfer characteristic gives the response of the DP4T inverter switch circuit with the antennas (ANT_1 and ANT_2)

2.9 RF Switch Performance Parameters

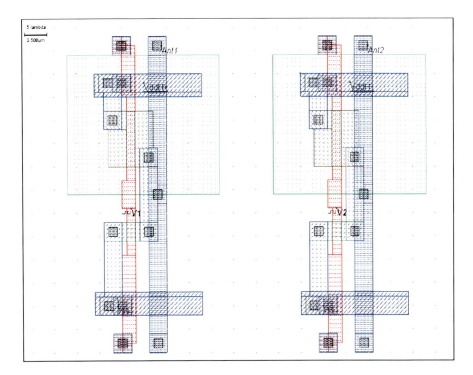

Fig. 2.8 Proposed DP4T switch layout with two transistors

and the specific input voltages V_1 and V_2. The gate-to-source voltage V_{gs} of the n-type MOSFET is equal to V_{in}, while the gate-to-source voltage of the p-type MOSFET is $V_{gs}^p = V_{in} - V_{dd}$, and the drain-to-source voltage of the p-type MOSFET is $V_{ds}^p = V_{ds}^n - V_{dd}$. From the output characteristics of the two transistors, the resulting drain currents in inverter circuit must be equal for each V_{in} and considering that the drain currents I_d of both the transistors must be equal, the voltage transfer characteristic is extracted from the layout of DP4T CMOS switch, as shown in Fig. 2.8. From this layout, it is obvious that when p-type MOSEFT is ON, then ANT_1 and ANT_2 are connected to the A_T_x and B_T_x, respectively, which are shown here with the 5 V or V_{dd}. Similarly when the n-type MOSEFT is ON, then ANT_1 and ANT_2 are connected to the A_R_x and B_R_x, respectively.

2.9 RF Switch Performance Parameters

The switches turn RF power ON and OFF, or perform high-frequency signal routing. The electrical parameters which have been measured and observed in this chapter exhibited by a switch are as follows.

2.9.1 Insertion Loss

The insertion loss is the loss of signal energy and power due to the insertion of a device in the transmission line (it may be optical fiber or LAN). The insertion loss of RF switch is the RF loss dissipated in the device and it is measured by S_{21} (in two-port device) between the input and output of the switch in its switch ON-state, which is the closed state for a series switch [48]. It may also be expressed as the reciprocal of the ratio of the signal power delivered to that part of the line following the device to the signal power delivered to that same part before insertion. The main contributing factors include resistive loss of the signal lines, contact at low to medium frequencies, and loss due to the skin depth effect where skin depth stands for the depth at which the electric current flows, measured from the surface of conductor.

2.9.2 Return Loss

The return loss is the loss of signal energy/power resulting from the reflection caused at a discontinuity in the transmission line (it may be optical fiber or LAN). This discontinuity can be a mismatch with the terminating load or with a device inserted in the line. The return loss of RF switch refers to the RF loss reflected back by the device means that portion of a signal that cannot be absorbed by the switch, or cannot cross an impedance mismatching due to the switch. This component of the signal is reflected from the impedance mismatching and returns back from that point and it is measured by S_{11} at the input of the switch in its switch ON-state. The main contributing factors include the mismatch of the switch's total characteristic impedance [48].

2.9.3 Isolation

The isolation of RF switch refers to the RF isolation between the input and output and it is measured by S_{21} of the switch in its blocking state, which is the OFF-state for a series switch. The main contributing factors include capacitive coupling and surface leakage [32].

2.9.4 RF Power Handling

This is a measure of how much and how well a switch passes the RF signal. To quantify RF power handling, the 1 dB compression point is commonly specified. That point is a measure of the deviation from the linearity of the 1 dB output power with respect to the input power. Alternatively, in pulsed power operation conditions, the peak pulsed power, the repetition rate, and the duty cycle are specified. In the switches containing P–N junctions, power handling is a function of frequency [8].

2.9.5 Linearity

If the ratio of a switch's output power to input power is a function of the input power level, then the switch is said to behave as a nonlinear device [9]. When signals of various frequencies are passed through the switch at a time, then in addition to the input frequencies, the switch's output will also contain frequencies related to the sum and difference of the harmonics of the various input frequencies. It is defined as input third-order intercept point (IIP_3) and output third-order intercept point (OIP_3).

2.9.6 Transition Time

Transition time is basically the time required for RF voltage to increase from 10 to 90 % (sometimes 0–100 %) for ON-state or decrease from 90 to 10 % (sometimes 100 to 0 %) for OFF-state.

2.9.7 Switching Speed

Switching speed is the time required for the switch to respond at the output upon applying of input voltage or changes in input voltage level. Switching speed includes drive propagation delay as well as transition time and is measured from the 50 % point on control voltage to 90 % for ON-state or 10 % for OFF-state of the RF voltage.

2.10 Topologies for DP4T Switches

At the high frequencies signal degradation occurs due to power dissipation in the line and power loss due to reflections in the transmission line. In a typical RF system, the switch network consists of multiple modules and cables. These networks may be constructed in various arrangements with modules of varying topologies. This topology is one of the most important features to consider when selecting an RF switch. Selecting a switch with the wrong topology can cause considerable effects on switching speed, propagation delay, insertion loss, and voltage standing wave ratio (VSWR).

With the growing degree of integration in the wireless communication systems, the test of RF front-end circuit is becoming more difficult. The high cost of test equipment and complexity of test procedure are becoming problems to be considered carefully by designers. In order to improve these problems, the idea of design for testability is introduced. In this case, the built-in self-test technology is implemented with an additional circuitry to the front-end chip. With this technology, the complete chip can be tested without external equipment. To test the switch circuit, two main types of topologies available for RF are multiplexers and SPDT relays [43].

a. A multiplexer is a switching system that sequentially routes various inputs to one output or vice versa. It is an ideal for reducing the channel count of RF devices such as analyzers and generators [47]. An application of multiplexer would be on the production floor to test the functionality of a mass produced RF device such as an integrated circuits or cell phones. In a case of batch of 1,000 device test, it would be cost prohibitive to dedicate an RF analyzer for the testing of each individual cell phone as this would require 1,000 RF instruments which increase the losses interns of power (energy) and delay in analyzing. Another way to conduct this test would be to manually route all 1,000 devices, one at a time, to a single RF instrument. This route would be more affordable than having multiple RF instruments. It reduces the test time significantly. The better test for this particular process would be to build an automated RF test system that uses a 1,000 × 1 multiplexer to route 1,000 devices to the RF analyzer which takes a measurement.
b. An SPDT relay is a basic form of multiplexer. This relay can route two inputs to one output or vice versa. General-purpose RF relay modules are usually made up of SPDT relays and are used to route a signal between two places [49]. For example, if a signal needs to be analyzed using an RF analyzer and then broadcasted to two locations through an antenna [50, 51], it can be routed to both locations using an SPDT relay.

The proposed DP4T switch can route four inputs to two outputs at a time or vice versa. So it is twice effective as compared to the existing SPDT switches as shown in Fig. 2.7.

2.11 Conclusions

The DP4T switch is designed with low insertion loss and low control voltage. The advantage of this switch is its minimum distortion and negligible voltage fluctuation, and it does not require large resistance at the receiving end. The proposed DP4T switches can be easily implemented into MIMO systems to increase the diversity and system capacity due to the multiple antenna usage. For the low-power circuits, a favorable condition can be achieved when both transistor gates are on the same potential contribution, even a reduced amount of leakage current. The proposed DP4T RF switch design with two parallel MOSFET modifies a conventional analog switch circuit design to operate with digital signals to achieve isolation buffering for bidirectional signals and high-density packing of multiple buffer switches operating under single enable control in a single package. However, certain modifications to a conventional analog switch design that facilitate the use of present switch in high-speed digital switching applications provide rapid switching times between ON-state and OFF-state, a low ON-resistance of less than 25 Ω, high-density configuration of multiple switches, and the ability to control the switches with standard logic level signals.

References

1. R. Granzner, S. Thiele, and Frank Schwierz, "Quantum effects on the gate capacitance of tri-gate SOI MOSFETs," *IEEE Trans. on Electron Devices*, vol. 57, no. 12, pp. 3231–3238, 2010.
2. N. Wang, "Transistor technologies for RFICs in wireless applications," *Microwave J.*, pp. 98–110, Feb. 1998.
3. Jack Browne, "More power per transistor translates into smaller amplifiers," *Microwaves and RF*, vol. 6, pp. 132–136, Jan. 2001.
4. Peerapong Uthansakul, Nattaphat Promsuwanna, and Monthippa Uthansakul, "Performance of antenna selection in MIMO system using channel reciprocity with measured data," *Int. J. of Antennas and Propagation*, vol. 2011, pp. 1–10, 2011.
5. S. Sanayei and N. Aria, "Antenna selection in MIMO systems," *IEEE Communications Magazine*, vol. 42, no. 10, pp. 68–73, Oct. 2004.
6. A. M. Street, *RF Switch Design*, IEE Training Course, United Kingdom, vol. 4, pp. 1–7, April 2000.
7. Kevin Walsh, *RF Switches Guide Signals in Smart Phones*, Skyworks Solutions Inc., Woburn, MA, Sept. 2010.
8. Application note, *Nanomount PIN diode switches data sheet*, AN 708, Microsemi Corporation, California, USA, 2006.
9. Application note, *Design with PIN Diodes*, APN 1002, Skyworks Solutions Inc., Woburn, MA, July 2005.
10. Application note, *GaAs T/R switch MMIC data sheet*, HMC223MS8, Hittite Microwave Corporation, Massachusetts, USA, Feb. 2001.
11. M. Steyaert, J. Janssens, B. Muer, M. Borremans, and N. Itoh, "A 2 V CMOS cellular transceiver front-end," *IEEE J. of Solid State Circuits*, vol. 35, no. 12, pp. 1895–1907, Dec. 2000.
12. S. Ahmed, C. Ringhofer, and D. Vasileska, "An effective potential approach to modeling 25 nm MOSFET devices," *J. of Computational Electronics*, vol. 9, no. 3-4, pp. 197–200, Oct. 2010
13. Joseph J. Carr, *Secrets of RF Circuit Designs*, 3rd Edition, Tata McGraw-Hill, India, 2004
14. R. Langevelde and F. Klaassen, "An explicit surface potential based MOSFET model for circuit simulation," *Solid State Electronics*, vol. 44, no. 3, pp. 409–418, March 2000.
15. Sungmo Kang and Yusuf Leblebichi, *CMOS Digital Integrated Circuits Analysis and Design*, 3rd Edition, McGraw-Hill, New York, USA, 2002.

16. Maria Villarroya, Eduard Figueras, and Nuria Barniol, "A platform for monolithic CMOS-MEMS integration on SOI wafers," *J. of Micromechanics and Microengineering*, vol. 16, no. 10, pp. 2203–2210, Oct. 2006.
17. Maria Villarroya and Nuria Barniol, "CMOS-SOI platform for monolithic integration of crystalline silicon MEMS," *Electronics Letters*, vol. 42, no. 14, pp. 800–801, July 2006.
18. J. Giner and Nuria Barniol, "VHF monolithically integrated CMOS-MEMS longitudinal bulk acoustic resonator," *Electronics Letters*, vol. 48, no. 9, pp. 514–516, April 2012.
19. F. J. Huang and O. Kenneth, "A 0.5 μm CMOS T/R switch for 900 MHz wireless applications," *IEEE J. of Solid State Circuits*, vol. 36, no. 3, pp. 486–492, March 2001.
20. T. Ohnakado, A. Furukawa, E. Taniguchi, and T. Oomori, "A 1.4 dB insertion loss, 5 GHz transmit/receive switch utilizing novel depletion layer extended transistors in 0.18 μm CMOS process," *Proc. of Symp. on VLSI Technology*, Honolulu, USA, 11-13 June 2002, pp. 162–163.
21. Lawrence E. Larson, "Integrated circuit technology options for RFICs present status and future directions," *IEEE J. of Solid State Circuits*, vol. 33, no. 3, pp. 387–399, March 1998.
22. D. Su, M. Zargari, P. Yue, D. Weber, B. Kaczynski, and B. Wooley, "A 5 GHz CMOS transceiver for IEEE 802.11a wireless LAN," *Proc. of IEEE Int. Conf. on Solid State Circuits*, San Francisco, California, USA, 7 Feb. 2002, pp. 92–93.
23. D. Su, M. Zargari, P. Yue, D. Weber, B. Kaczynski, and B. Wooley, "A 5 GHz CMOS transceiver for IEEE 802.11a wireless LAN systems," *IEEE J. of Solid State Circuits*, vol. 37, no. 12, pp. 1688–1694, 2002.
24. M. Uzunkol and G. M. Rebeiz, "A low loss 50-70 GHz SPDT switch in 90 nm CMOS," *IEEE J. of Solid State Circuits*, vol. 45, no. 10, pp. 2003–2007, Oct. 2010.
25. Viranjay M. Srivastava, K. S. Yadav, and G. Singh, "DP4T RF CMOS switch: A better option to replace SPDT switch and DPDT switch," *Recent Patents on Electrical and Electronic Engineering*, vol. 5, no. 3, pp. 244–248, Oct. 2012.
26. C. Lee, B. Banerjee, and J. Laskar, "Novel T/R switch architectures for MIMO applications," *IEEE Microwave Symp. Digest*, vol. 2, pp. 1137–1140, June 2004.
27. N. Talwalkar, C. Patrick Yue, and S. Simon Wong, "Integrated CMOS transmit-receive switch using LC tuned substrate bias for 2.4 GHz and 5.2 GHz applications," *IEEE J. of Solid State Circuits*, vol. 39, no. 6, pp. 863–870, June 1989.
28. J. P. Carmo, P. M. Mendes, C. Couto, and J. H. Correia, "A 2.4 GHz RF CMOS transceiver for wireless sensor applications," *Proc. of Int. Conf. on Electrical Engineering*, Coimbra, 2005, pp. 902–905.
29. Piya Mekanand and Duangrat Eungdamorang, "DP4T CMOS switch in a transciever of MIMO system," *Proc. of 11th IEEE Int. Conf. of Advanced Communication Technology*, Korea, 15–18 Feb. 2009, pp. 472–474.
30. P. H. Woerlee, M. J. Knitel, and A. J. Scholten, "RF CMOS performance trends," *IEEE Trans. on Electron Devices*, vol. 48, no. 8, pp. 1776–1782, Aug. 2001.
31. Neil H. E. Weste and Kamran Eshraghian, *Principles of CMOS VLSI design: A system perspective*, 2nd Edition, Addison Wesley, USA, 2005.
32. J. P. Carmo, P. M. Mendes, C. Couto, and J. H. Correia, "A 2.4 GHz wireless sensor network for smart electronic shirts integration," *Proc. of IEEE Int. Symp. on Industrial Electronics*, Vigo, Spain, 4–7 June 2007, pp. 1356–1359.
33. Chien Cheng Wei, Hsien Chin Chiu, Shao Wei Lin, Ting Huei Chen, Jeffrey S. Fu, and Feng Tso Chien, "A comparison study of CMOS T/R switches using gate/source terminated field plate transistors," *Microelectronic Engineering*, vol. 87, no. 2, pp. 225–229, Feb. 2010.
34. C. Tinella, J. Fournier, D. Belot, and V. Knopik, "A high performance CMOS SOI antenna switch for the 2.5 GHz to 5 GHz band," *IEEE J. of Solid State Circuits*, vol. 38, no. 7, pp. 1279–1283, July 2003.
35. Mei Chao Yeh, Zuo Min Tsai, and Chih Ping Chao, "A millimeter-wave wideband SPDT switch with traveling-wave concept using 0.13-μm CMOS process," *IEEE Int. Microwave Symposium Digest*, pp. 4, 12–17 June 2005.

References

36. T. Quemerais, L. Moquillon, J. Fournier, and P. Benech, "A SPDT switch in a standard 45-nm CMOS process for 94 GHz applications," *Proc. of European Microwave Conf.*, Paris, France, 26 Sept.–1 Oct. 2010, pp. 425–428.
37. T. Dinc, S. Zihir, and Y. Gurbuz, "CMOS SPDT T/R switch for X-band, on-chip radar applications," *IET Electronics Letters*, vol. 46, no. 20, pp. 1382–1384, 2010.
38. Chen Lei, Tian Liang, Zhou Jin, Huang Ai bo, and Lai Zongsheng, "A high performance PD SOI CMOS single-pole double-throw T/R switch for 2.4 GHz wireless applications," *Proc. of 5th Int. Conf. on Wireless Communications, Networking and Mobile Computing*, Beijing, China, 24–26 Sept. 2009, pp. 1–4.
39. Siti Maisurah, S. Rasidah, Abdul Rahim, and Y. M. Razman, "A 0.18 μm CMOS T/R switch for 900 MHz wireless application," *Proc. of IEEE Int. Conf. on RF and Microwave*, Kuala Lumpur, Malaysia, 2–4 Dec. 2008, pp. 176–179.
40. Mei Chao Yeh, Zuo Min Tsai, and Huei Wang, "A miniature DC to 50 GHz CMOS SPDT distributed switch," *Proc. of European Symp. on Gallium Arsenide and Other Semiconductor Application*, Paris, France, 3–4 Oct. 2005, pp. 665–668.
41. S. F. Chao, H. Wang, C. Y. Su, and J. G. Chern, "A 50 to 94 GHz CMOS SPDT switch using traveling wave concept," *IEEE Microwave and Wireless Components Letters*, vol. 17, no. 2, pp. 130–132, Feb. 2007.
42. Viranjay M. Srivastava, K. S. Yadav, and G. Singh, "Capacitive model and S-parameters of double-pole four-throw double-gate RF CMOS switch," *Int. J. of Wireless Engineering and Technology*, vol. 2, no. 1, pp. 15–22, Jan. 2011.
43. Application Note and Product catalogue, *RF switch performance advantages of ultra CMOS technology over GaAs technology*, AN 18, Peregrine Semiconductor, USA.
44. Joe Grimm, *CMOS RFIC switches: Simple and inexpensive, the latest 2.5 GHz versions pose a legitimate challenge to GaAs switches*, A Product Catalogue, RFIC Switches, California Eastern Laboratories, Santa Clara, CA, USA, Jan. 2004.
45. S. H. Lee, C. S. Kim, and H. K. Yu, "A small signal RF model and its parameter extraction for substrate effects in RF MOSFETs," *IEEE Trans. on Electron Devices*, vol. 48, no. 7, pp. 1374–1379, July 2001.
46. Yuan Taur, and Tak H. Ning, *Fundamentals of Modern VLSI Devices*, 1st Edition, Cambridge University Press, United Kingdom, 2008.
47. Viranjay M. Srivastava, K. S. Yadav, and G. Singh, "Analysis of drain current and switching speed for SPDT switch and DPDT switch with the proposed DP4T RF CMOS switch," *J. of Circuits, Systems and Computers*, vol. 21, no. 4, pp. 1–18, June 2012.
48. Sieu Ha, You Zhou, and P. Treadway, "Electrical switching dynamics and broadband microwave characteristics of VO2 radio frequency devices," *J. of Applied Physics*, vol. 113, no. 18, pp. 184501–184507, May 2013.
49. Mei Yeh, Zuo Tsai, and Ying Chang, "Design and analysis for a miniature CMOS SPDT switch using body floating technique to improve power performance," *IEEE Trans. on Microwave Theory and Techniques*, vol. 54, no. 1, pp. 31–39, Jan. 2006.
50. Viranjay M. Srivastava, K. S. Yadav, and G. Singh, "Design and performance analysis of double-gate MOSFET over single-gate MOSFET for RF switch," *Microelectronics Journal*, vol. 42, no. 3, pp. 527–534, March 2011.
51. Viranjay M. Srivastava, K. S. Yadav, and G. Singh, "Analysis of double gate CMOS for double-pole four-throw RF switch design at 45-nm technology," *J. of Computational Electronics*, vol. 10, no. 1–2, pp. 229–240, June 2011.

Chapter 3
Design of Double-Gate MOSFET

3.1 Introduction

Recent progress to scale down the transistors to smaller dimensions provides the faster transistors, as well as lowers the effective density in terms of transistors area. The transistor scaling necessitates the integration of new device structures. The Double-Gate (DG) MOSFETs are example of this, which are capable for nanoscale integrated circuits due to their enhanced scalability, compared to the bulk or Si-CMOS [1–5]. However, the better scalability can be achieved by introduction of a second gate at the other side of the body of transistor resulting in the double-gate structure. Due to excellent control of the short channel effects (SCE), double-gate devices have emerged as the device of choice for circuit design in sub-50 nm and below regime [6]. The low subthreshold leakage and higher ON-current in double-gate devices make them suitable for circuit design in sub-50 nm regime [7–10]. However, isolated or independent gate option can be useful for low power and mixed signal applications [11–15]. The double-gate devices can be classified on the basis of their structure. The front and back gates of double-gate devices are connected together resulting in a 3-Terminal device as discussed in Chap. 2. The 3-Terminal devices can be used for direct replacement of conventional single-gate bulk-CMOS devices. Recently, the double-gate devices with independent gate control option (separate contacts for back and front gates) have been developed such double-gate devices are referred to as isolated or independent gate devices as shown in Fig. 3.1. The isolated gate devices with a second gate for each device are referred to as 4-Terminal devices. In such technologies, one can choose to connect the back and front gates together or to control them separately while designing a circuit resulting in new circuit technology.

The first MOS devices truly had metal-gate electrodes as Au, Cr, or Al. Due to its ease of deposition and etching, adherence to SiO_2 and Si surfaces, and its freedom from corrosion. However, Al became the standard metal-gate electrode for early MOS devices. The important limitations such as electro-migration and spiking into shallow junctions have been overcome by alloying with Cu or Si. The large number

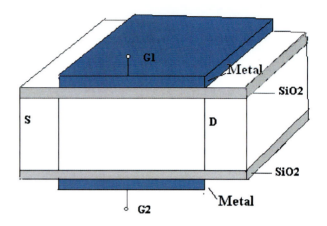

Fig. 3.1 Schematic of the basic n-type double-gate MOSFET

of gates provides improved electrostatic control of the channel, so that the Silicon body thickness and width can be larger than the ultrathin body silicon-on-insulator (SOI) and DG MOSFET structures, respectively. The gate electrodes are formed with a single deposited gate layer and are defined lithographically [16–19]. They are tied together electrically and are self-aligned with each other as well as the source/drain regions [20–24]. The challenge is slightly poorer in the electrostatic integrity than with double-gate structures, particularly in the corner regions of the channel [25–27]. The DG MOSFET significantly suppressed the short channel effects, reduced the drain-induced barrier lowering, excellent scalability, and has been regarded as a possible candidate for the device scaling at the end of the International Technology Roadmap for Semiconductors (ITRS) guidelines [28–32]. The double-gate devices exploit the volume inversion phenomenon where the carriers flow inside the silicon film thereby improving the carrier mobility [33–35]. DG MOSFETs can be fabricated with its channel in the plane of the wafer, as in the standard configuration, where recorded currents have been achieved in test structures [36–40]. As sub-25 nm gate length DG MOSFETs have been undoped ultrathin Silicon film for enhanced channel mobility. It is a possible option to control SCEs and achieve low OFF-current (I_{off}) for the source/drain extension regions by optimizing lateral straggle and spacer width, instead of the conventional method of increasing the channel doping and altering the thickness of the silicon film. This concept of non-overlapped gate-source and gate-drain extension regions (also known as gate underlap) has been proposed for a sub-20 nm bulk MOSFET [41–44]. For the nanoscale double-gate devices, the intrinsic channel avoids random dopant fluctuations [45] and offers higher carrier mobility. It also indicates the possibility of a source/drain punch through when the source/drain extension regions are heavily doped to minimize the extension region resistance to achieve higher ON-current (I_{on}). Most of the work on the double-gate devices has mainly focused on the fabrication and characterization [46, 47], carrier mobility [48, 49], analog/RF applications [50], circuits [51, 52], sensitivity analysis [53], gate misalignment [54], and modeling [55, 56] for the ideal devices with abrupt source/drain.

3.1 Introduction

Some literatures have been reported on the source/drain extension region design and optimization for the single-gate devices [57], double-gate MOSFETs [58–60], and vertical FinFETs [61, 62] for digital and analog communication applications.

The symmetric DG MOSFET structure has been emphasized for the application of RF switch due to its intrinsic strength to the short channel effects (SCE) and it improves the current drive capability. When we are using a switch with multiple gates, the behavior of these switches depends on the number of gates, which controls the operational process of the device. Therefore, the additional logic functions can be implemented into a single chip transistor. The transistor which uses the independently controlled gates are not limited to only two gates, but due to the geometrical reasons of the transistor and the connectivity of the transistor terminals, it is suitable to use only two gates. The independent double-gate transistors can be used to implement the universal logic functionality within a single transistor [63].

Earlier, Gidon [64] has investigated the 2-dimensional DG MOSFET and combat the high aspect ratio of the transistor (thin channel compared to its length) by introducing an anisotropy scale factor in its geometry. Lu and Taur [65, 66] have presented an analytic potential model for long channel symmetric and asymmetric DG MOSFETs. The model is derived rigorously from the exact solution to the Poisson's continuity equation and current continuity equation without the charge-sheet approximation. To preserve the proper physics, volume inversion in the subthreshold region is well accounted for the model proposed by Fossum [67]. The analytical expressions of the drain current, terminal charges, and capacitances, long channel DG MOSFETs are found continuous in all operating regions, such as linear, saturation, and subthreshold. The drain current model, charge model, transconductance model, and capacitive model for symmetrical and asymmetrical DG MOSFETs are also developed by Fossum [68]. However, the proposed switch experiences the minimal distortion, negligible voltage fluctuation, and low voltage power supply. A better conformity between the numerical simulations and analysis of the proposed model has been achieved [69]. For the aspect of RF switch, the bandwidth depends on the capacitance connected to the ground, which is due to the sidewall capacitances present in the MOSFET structure [70, 71]. The logic density of a transistor can be increased with independently controlled DG MOSFET [72].

The scalability of such a device structure is limited due to increased SCE [73], which has motivated the need for nonclassical Silicon devices to extend CMOS scaling beyond the 45-nm node. However, the ultrathin body SOI FETs employ very thin Silicon body to achieve better control on the channel by the gate, and hence, reduced the leakage current and short channel effects. The use of intrinsic or lightly doped body reduces the threshold voltage (V_{th}) variations due to random dopant fluctuations [74]. It enhances the mobility of the carriers in the channel region and therefore switch ON-current [75, 76]. Due to the excellent control of SCE, low subthreshold leakage and higher ON-current in the double-gate devices make them suitable for circuit design in sub 50-nm regime [77, 78]. The double-gate devices with isolated gates (independent gates) are also being developed [79].

However, the independent gate option can be useful for low power and mixed signal applications [80, 81] and such developments at the device level provide opportunities for new ways of circuit design for the low power and high performance devices.

3.2 Design Process of Double-Gate MOSFET

The Polycrystalline Silicon (poly-Si) is used for a gate material of MOSFET due to its high compatibility with fabrication process. This gate material can be very easily deposited by low pressure chemical vapor deposition techniques and having better thermal stability on SiO_2, which works as the gate dielectric for MOSFETs. A useful property of the gate electrode is its work function at the dielectric interface, which controls the threshold voltage. Due to the scaled down over successive technology generations the dopant concentrations in the channel have been increased to maintain better short channel performance, and to prevent the channel depletion region. A depletion layer in the gate introduces a capacitance in series to the gate dielectric capacitance and thus adds to the effective dielectric thickness between the gate and the channel, which implies the reduction in capacitive coupling between the gate and the channel in inversion. For comparable channel currents, the reduction in the inversion capacitance in the case of a poly-Si gate device lowers the drive current significantly.

The DG MOSFET is a natural extension of SOI devices. The double gate gives rise to various performances such as increased transconductance and a lower threshold voltage. For symmetrical type, the thickness of back oxide layer is identical as of front oxide and identical gate materials are used, which allows both the gates to control the operation of the device. For asymmetrical type, unlike oxide thickness may be used and materials of different work function as n^+ poly and p^+ poly can be used for the front and back gate. Since with the symmetric-gate design, the channel area is raised to increase the saturation current and the Silicon body control is enhanced to reduce the short channel effects. Recently, the development of ultrathin DG MOSFET introduces the concept of volume inversion [28]. The inversion charge spreads throughout the ultrathin Silicon body, which improves the device characteristics (as higher current due to the substrate mobility) and strong inversion provides the information of the swept charge as well as the saturation current in MOSFET device physics [82, 83].

In DG MOSFET, the second gate is added opposite to the traditional gate have long been recognized [84, 85] for their potential to better control of the short channel effects. The short channel effects limit the minimum channel length at which the MOSFET device is electrically well functioned. As the channel length of MOSFET is reduced, the drain potential begins to strongly influence the channel potential. This short channel effect is mitigated by the use of thin gate oxide and thin depletion width below the channel to the substrate, to shield the channel from the drain. Further the decrease of the depletion region degrades the gate influence

3.2 Design Process of Double-Gate MOSFET

on the channel and leads to a slower turn ON of the channel region. In DG MOSFETs, the longitudinal electric field generated by the drain is better screened from the source end of the channel due to the proximity to the channel of the second gate, result the reduced short channel effects, reduced drain-induced barrier lowering and improve the subthreshold swing. Therefore, as MOSFET scaling becomes limited by leakage currents, DG MOSFET offers the opportunity to proceed beyond the performance of single-gate bulk-Silicon MOSFET. From a bulk-Silicon device design perspective, the increased body doping concentration could be employed to reduce the drain-induced barrier lowering (DIBL). It would also increase the subthreshold swing, thereby requiring higher threshold voltage (V_{th}) to keep the subthreshold current adequately low. The drain-induced barrier lowering [86] is a well-publicized short channel effect in scaled MOSFETs. In accordance with the 2-dimensional Poisson equation in the weakly inverted channel/body, high drain bias (V_{ds}) lowers the potential barrier height at the virtual source, which allows increased carrier diffusion from the source to the channel, causing higher OFF-current, and lowering the (saturation) threshold voltage. By decreasing the body doping concentration the subthreshold swing can be improved; however it degrades DIBL due to this reason and a compromise is necessary for the bulk-Silicon device design.

In DG MOSFET, when voltage is applied to the gates of device, the active Silicon region is so thick, that the control region of the Silicon remains controlled by the majority carriers in the region. This causes not one but two channels to be formed. One channel forms near the top boundary between Silicon and the Silicon insulator and the other one forms likewise at the bottom interface as a DG MOSFETs structure is shown in Fig. 3.1. These two channels are separated by sufficient distance as to be independent of each other, which creates two independent transistors on the same piece of Silicon. Each gate can control one-half of the devices and its operation is completely independent to each other. The total current through the device is equal to the sum of the currents through the separate channels. The relative scaling advantage of the DG MOSFET is about twofold. The performance of the symmetrical version of the DG MOSFET is further increased by higher channel mobility compared to a bulk MOSFET. Since the average electric field in the channel is lower, which reduces the interface roughness scattering according to the universal mobility model [72, 87, 88]. Here the two main device processes are possible for double-gate devices, namely:

a. Symmetric device with same gate material and oxide thickness for the front and the back gates [18, 89].
b. Asymmetric device with different strengths for front and back gates.

However, various strengths can be obtained by using either different oxide thickness (asymmetric oxide) [90] or materials of different work function (for example, n$^+$ poly and p$^+$ poly) for the front and back gate (asymmetric work function) [91]. Independent control of front and back gate in double-gate devices can be effectively used to improve the performance and reduce the power in sub 50-nm circuits. The independent gate control can be used to merge the parallel transistors in the noncritical paths which results the reduction in effective switching capacitance and hence the power dissipation. A variety of examples for the independent gate operation of the double-gate devices are dynamic logic circuits, Schmitt triggers, sense amplifiers, and static random access memory (SRAM) cells. In addition to the independent gate option, we also discussed the usefulness of asymmetric devices and the impact of width quantization and process variations on circuit design.

When a proper model and a set of model parameters are used then we can obtain accurate and useful circuit. Thus, it is imperative to develop a well-constructed model and its parameter extraction method in order to effectively describe and predict the device characteristics. A model is normally represented by an equivalent circuit consisting of elements such as resistances, capacitances, inductances, voltage sources, current sources, and heat sink. These elements are described by mathematical equations. All of the constants and coefficients associated with the equations, called the model parameters. However, some of them may result from empirical experiences to enhance the model accuracy. The model parameters need to be determined from extraction techniques based on the experimental data from MOSFETs measured at different biased condition and frequencies. At low frequencies, these measurements are usually performed by utilizing Z, Y, or H matrices. It is easier to measure the voltage or current with open or short circuits however these properties are suitable for the low frequencies. By terminating the port with a cable of characteristic impedance, the S-parameter technique measures the power waves propagating into and being reflected from the device and thus is the easiest and the most reliable way to characterize the high-frequency networks.

3.3 Effects of Double-Gate MOSFET on the Leakage Currents

In double-gate structures, the presence of two gates and ultrathin body helps to reduce the SCE, which significantly reduces the subthreshold leakage current [92]. However, lower SCE in the double-gate devices and higher the drive current (due to two gates) allow the use of thicker oxide in the double-gate devices compared to the bulk-CMOS structures, which helps to reduce the gate leakage current. The lower SCE allows the use of lower body doping (body can even be intrinsic) in the double-gate devices compared to the bulk-CMOS structure. Hence, to induce the equal inversion charge, the double-gate devices required lower electric field compared to the bulk double-gate structure, which also helps to reduce the gate leakage current in the double-gate devices [93]. Although the leakage current is significantly reduced in double-gate devices, however, it is important to analyze different leakage current mechanisms in such devices. In particular, different double-gate device options, principally, symmetric doped body poly-gate devices (SymDG), symmetric intrinsic

3.3 Effects of Double-Gate MOSFET on the Leakage Currents   51

body mid-gap devices (MGDG), and asymmetric (n^+ poly/p^+ poly) intrinsic body (AsymDG) devices [94], have a strong impact on the different leakage components. Hence, it is necessary to analyze the double-gate device structure, which is most suitable for the low leakage circuit design. One of the major advantages of using DG MOSFET is the lower leakage current and smaller subthreshold voltage [95]. The major leakage components in the double-gate devices are given below.

3.3.1 Subthreshold Leakage

The electrically coupled front, back gates and ultrathin body reduce the short channel effect in the double-gate devices, consequently the reduction of subthreshold leakage [96, 97]. For an equal "ON" current of multi-gate (MGDG) device shows the lower subthreshold leakage compared to the SymDG (symmetrical) and AsymDG (asymmetrical) devices, due to the fact that the surface electric field is higher in poly-gate devices (due to higher doping) and asymmetric devices (at the front gate due to large work-function difference between front and back gates) which reduces the mobility (due to higher surface scattering).

3.3.2 Gate Leakage

The Gate leakage in double-gate devices is due to the gate-to-channel tunneling and overlap tunneling current because of the presence of lower body doping (as SCE is controlled by two gates and ultrathin body) oxide-field is lower in the double-gate devices as compared to the bulk devices (for equal inversion charge) [93]. Hence, the double-gate devices show the lower gate-to-channel tunneling as compared to the bulk devices. This effect is obvious in MGDG devices as the body is intrinsic. So, the mid-gap intrinsic body devices have lower gate-to-channel leakage compared to SymDG devices. In AsymDG devices due to the work-function difference of the front and back gates, front surface field is very high whereas back surface field is negligible. Hence, the tunneling occurs only through the front gate (which is larger than MGDG device but smaller than the SymDG device).

3.3.3 Band to Band Tunneling of Electrons

Band to band tunneling (BTBT) across the reverse-biased p-n junction from the p-side valence band to the n-side conduction band is becoming an important component of the leakage in nanoscale bulk and SOI CMOS technologies. It is manifested both as gate-induced drain leakage (GIDL) in the drain-gate overlap region and reverse-biased junction leakage in the halo-implant region. With DG

CMOS likely to become the mainstream technology around the 45 or 32-nm node and it is important to study the BTBT phenomenon in these devices. A detailed study and modeling of BTBT leakage for the double-gate devices can be obtained in [98]. The device level parameters that impact power and performance include the geometrical parameters and the electrical properties of the materials used for gate (work function). In particular, it is important to analyze the impact of gate electrode thickness and gate underlap on the fringe capacitance of nanoscale DG MOSFET.

However, Moldovan et al. [99] have demonstrated the capability of undoped DG MOSFET with explicit and analytical compact model to forecast the effect of the volume inversion on the intrinsic capacitances. For this purpose, Moldovan et al. [99] also simulated the results for these capacitances, which present an accurate dependence on the Silicon layer thickness, consistent with 2-D numerical simulations, for both thin and thick Si-films. Also, if the current drive and transconductance are enhanced in volume inversion regime, this compact model results that intrinsic capacitances are higher as well, which may limit the high-speed (delay time) behavior of DG MOSFETs under the volume inversion regime. Hamed et al. [100] emphasized the use of independently driven nanoscale DG MOSFETs for the low power analog circuits and suggested that in the independent drive configuration, the top-gate response of DG MOSFETs can be altered by application of a control voltage on the bottom gate. The extremely thin film in the DG MOSFET, which uses the volume inversion concept, has been formulated by Moldovan et al. [99] and Balestra and Elewa [101] in detail. This device has received significant attention from the perspective of their technological feasibility and theoretical modeling. However, the numerous advancement for the device architecture have been explored as gate-all-around (GAA), delta, lateral epitaxial overgrowth, folded-gate, fin-gate, self-alignment [102] are few. The electrostatic and Monte-Carlo simulations established the detailed advantage of the DG MOSFETs as discussed in refs. [103, 104]. An impressive compact and analytical model for the DG MOSFETs, which account for the quantum, volume inversion, short channel, and non-static effects have been proposed by Ge and Fossum [105]. The threshold voltage increases gradually in SOI films due to the incidence of the quantum effects. The small discontinuity is observed around the second gate voltage value of zero ($V_{G_2} = 0$), due to the formation of a depletion region, underneath the buried oxide, which increases the apparent buried oxide thickness [106]. The quantum mechanical effects have not been considered in this model because it is negligible for Si-films thicker than 10 nm. For films thinner than 10 nm, the quantum confinement should be considered. It leads to a reduction of the channel charge density and an increase of the threshold voltage [107].

In the thicker and fully depleted MOSFETs, the drain current undershoot can be observed when the front gate is biased in inversion and the back gate is suddenly switched from depletion to accumulation. The advantage of extremely thin devices is that they do not suffer from such transients because the back interface cannot be driven in accumulation. Since the primary intrinsic variations include the random dopant fluctuation, which is caused by the uncertainty in charge location and charge numbers, such as discrete placement of dopant atoms in the channel region that follow a Poisson distribution [108]. As the device size scales down, the total number of channel dopants decreases, which provides a larger variation of dopant numbers, and significantly impacting threshold voltage. Dollfus and Retailleau [109] have compared the noise performances of DG MOSFET and single-gate (SG) MOSFETs and discussed the noise-figure in the structures, that is explained in terms of a favorable increase of cross-correlation between the drain and gate currents. The authors also proved that the presence of a residual undesired charged impurity in the channel of a double-gate structure induces perceptible changes in the spectral density of the gate current (I_g) fluctuations that modifies the noise-figure [109, 110]. A comprehensive analysis to limit the short channel effects in SG and DG SOI/GOI MOSFETs, based on the ratio of effective channel length to natural length, has been presented by Sharma and Kumar [111]. These results suggest that Germanium–on–insulator (GOI)-based devices exhibit higher degree of SCE as compared to SOI-based MOSFETs [112, 113].

In general for the symmetrical double-gate device, the front gate remains relatively constant with back-gate bias, as the gate-to-gate coupling is limited at the strong inversion charge at back surface. In the similar way for asymmetrical double-gate device, the modulation effect is very significant, as the gate-to-gate coupling is extended until the back surface becomes strongly inverted (the weak back-channel has a high, about 1.0 V higher than the front-channel). In bulk and partially depleted (PD)/SOI devices, the effectiveness and operating frequency of the well/body bias are limited by the distributed resistance and capacitance (R and C) of the well and body contact. It also tends to degrade with technology scaling due to the lower body factor in the scaled devices. The area compactness can be easily implicit in complex gates. The use of double gate increases the cell area compared to that of the same width single-gate device.

3.4 Performance Improvement of DG MOSFET over SG MOSFET

We have analyzed and discussed a DP4T RF switch using DG CMOS technology. In this technology the transistor with a gate-controlled bulk current using either an n-type or p-type substrate for the complementary transistor types are used such DG MOSFETs rely on majority carrier flow through the bulk of the source-drain Silicon passage. The carrier concentration in the central part of this passage (slit) is controlled by the potentials of two gates (G_1 and G_2). In other words, the changes of the bias of the gate junctions result in variations of penetration of the depletion layers into the substrate and modulate resistance of the channel.

Figure 3.2a shows the layout of n-type MOSFET with two symmetric-gate voltages and output voltage through V_{out} as DG MOSFET and Fig. 3.2b shows the SG MOSFET, respectively. Here the color codes have their usual meanings [114], which are used in the rest of the book. The layout is drawn with available tools for high-speed with the MOSFET width of 600 nm and length of 120 nm. These designs have a poly, drain, and source. The resistances are also present in this

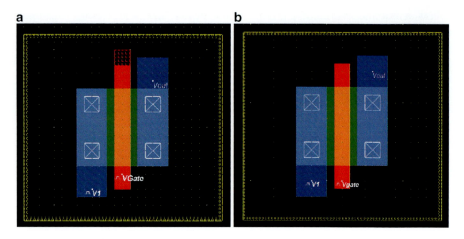

Fig. 3.2 Layout of (**a**) DG MOSFET and (**b**) SG MOSFET

layout due to the metal connection with input voltage and output voltage. This drain and source has equal capacitance of 0.19 fF, resistance of 90 Ω, and thickness of 2 μm, with metal capacitance of 0.13 fF, diffusion capacitance 0.06 fF, capacitance of the gate is 0.86 fF. DG MOSFET has a resistance of 68 Ω and thickness of 3 μm, whereas SG MOSFET has resistance of 32 Ω and thickness of 2 μm.

Here, we have analyzed the performance of DG MOSFET and SG MOSFET by applying a gate voltage 1.2 V as low level. The start time, rise time, fall time, and pulse time for this signal are taken as 0.475 ns, 0.025 ns, 0.025 ns, and 0.475 ns, respectively. In this simulation V_1 is 1.2 V as high level and start time, rise time, fall time, and pulse time for this signal are taken as 0.600 ns, 0.025 ns, 0.025 ns and 0.475 ns, respectively.

Assuming that DG MOSFET has symmetric-gate structure and voltage applied on both gates is same, Fig. 3.3a depicts that the output voltage (V_{out}) for DG MOSFET is high when both the drain voltage (V_1) and gate voltage (V_{gate}) are high, which is for duration 1.0–1.1 ns and we simulated the output voltage (V_{out}) 0.30 V and from Fig. 3.3b for SG MOSFET the simulated result, we obtained the output voltage (V_{out}) 0.90 V for the same duration. To determine the drain current, a conventional technique in thin-oxide MOSFETs consists of C–V measurements [115]. However, this DG MOSFET device has slightly thick oxides (so that a very small capacitance created).

Hence, the conventional charge equation $Q = C_{ox} \cdot (V_{gs} - V_{th})$ will be suitable, which yields direct and accurate values for the density of charge carriers, even with double activated gates, the linear relationship is perfectly preserved. By this charge Q, we can obtain the drain current using following equation:

$$I_{ds} = \mu \cdot Q \cdot V_{ds} \cdot \frac{W}{L} \qquad (3.1)$$

where μ and V_{ds} are the channel mobility and applied drain to source voltage, respectively. W and L are the channel length and width, respectively. Since in DG

3.4 Performance Improvement of DG MOSFET over SG MOSFET 55

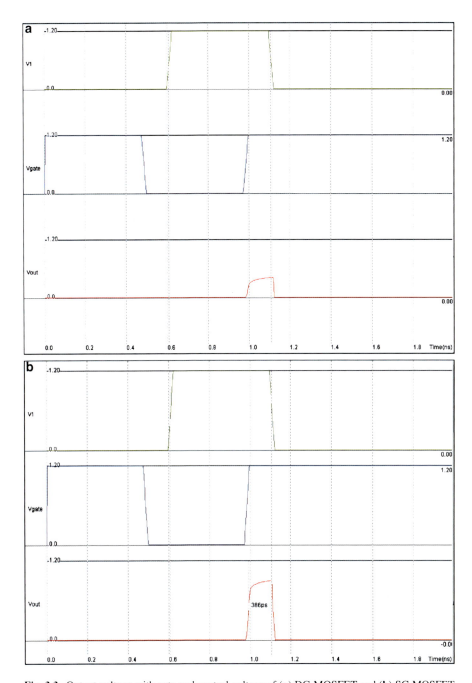

Fig. 3.3 Output voltage with gate and control voltage of (**a**) DG MOSFET and (**b**) SG MOSFET

MOSFET charge Q is greater as compared to the SG MOSFET, due to the higher capacitance values (as shown in the next section of this chapter), so the drain current is higher in DG MOSFET devices. For the duration of 1.0–1.1 ns, currents are drawn on log scale as shown in Fig. 3.4a for DG MOSFET, which is 80 µA and becomes stable at 0.10 µA and for SG MOSFET in Fig. 3.4b, which is 76 µA and becomes stable at 0.10 µA.

The impact of metal-gate work function on the threshold voltage and therefore on the leakage current (I_{OFF}) can be determined for the DG MOSFET. When the metal-gate work function is raised, I_{OFF} decreases extensively and threshold voltage increases [116]. In order to maintain I_{OFF} very low, it is necessary to increase the metal work function or the device resistance at the OFF-condition (R_{OFF}) should be very high as well as resistance at the ON-condition (R_{ON}) should be low (this is shown in the next section of this chapter). In addition to this, the increase in metal work function is accompanied by an increase in threshold voltage. After this threshold voltage is achieved, the output voltage can be found. This output voltage stabilizes at 0.53 V for the drain voltage (V_1) 0.60 V onwards for DG MOSFET as shown in Fig. 3.5a and for SG MOSFET this output voltage stabilizes at 1.03 V for the drain voltage (V_1) 1.10 V onwards as shown in Fig. 3.5b. So, we conclude from these parameters that the output voltage stabilization for DG MOSFET is less as compared to the SG MOSFET.

The layout for n-type DG MOSFET is shown in Fig. 3.6a and p-type DG MOSFET is shown in Fig. 3.6b. The scalable CMOS rules support both n-well and p-well processes. In Fig. 3.6b the green boundary shows the well for the designing of p-type DG MOSFET and also an extra V_{dd} is applied to support that well.

3.5 Resistive and Capacitive Model of DG MOSFET and SG MOSFET

The resistive and capacitive models of a MOSFET transistor which is biased in linear region, at the ON-state of switch, for DG MOSFET and SG MOSFET are shown in Fig. 3.7a, b respectively. For the given design of DG MOSFET and SG MOSFET under the operating condition, insertion loss is conquered by its ON-state resistance and substrate resistance [117, 118]. The isolation of the switch is finite due to the signal coupling through parasitic capacitances and junction capacitances. In the cut-off region, the MOSFET resistance R_{ON}, R_{ON1}, R_{ON2} will become zero. For maximum capacitance (as a worst case), assuming all the capacitances are present at a time. In DG MOSFET, parasitic capacitances are C_{ds1}, C_{ds2}, C_{gs1}, C_{gs2}, C_{gd1}, and C_{gd2} and junction capacitances are not present as bulk and are not available in this MOSFET whereas in SG MOSFET available parasitic capacitances are C_{ds}, C_{gs}, and C_{gd} and junction capacitances are C_{sb} and C_{db}.

3.5 Resistive and Capacitive Model of DG MOSFET and SG MOSFET   57

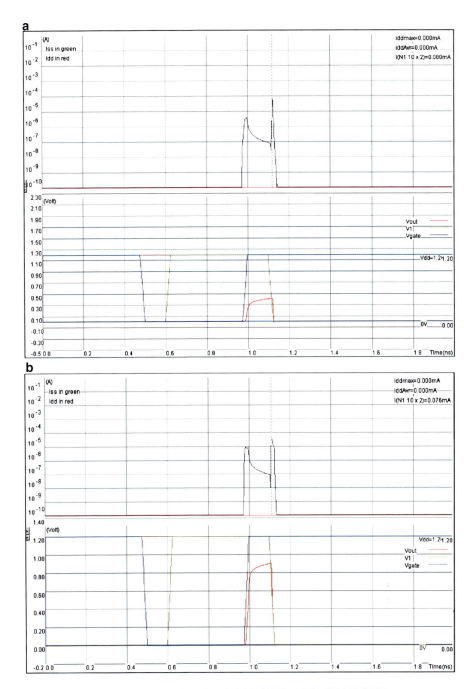

Fig. 3.4 Drain current characteristics of (**a**) DG MOSFET and (**b**) SG MOSFET

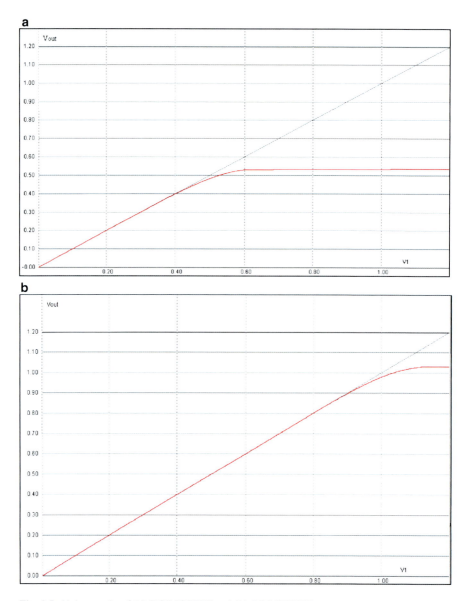

Fig. 3.5 Voltage gain of (**a**) DG MOSFET and (**b**) SG MOSFET

For DG MOSFET when both the transistors are ON, C_{sb} and C_{db} are not present, so fewer signals being coupled to the substrate as substrate is not present in this structure, so no dissipation in the substrate/bulk resistance (R_b). When the transistor is in cut-off region, increasing C_{ds1}, C_{ds2}, C_{gd1}, C_{gd2}, C_{gs1}, and C_{gs2} leads to higher isolation between the source and drain, due to no capacitive coupling between these terminals. Whereas for SG MOSFET, when the transistor is ON, increasing C_{sb} and C_{db} leads to more signal being coupled to the substrate/bulk and dissipated in the

3.5 Resistive and Capacitive Model of DG MOSFET and SG MOSFET 59

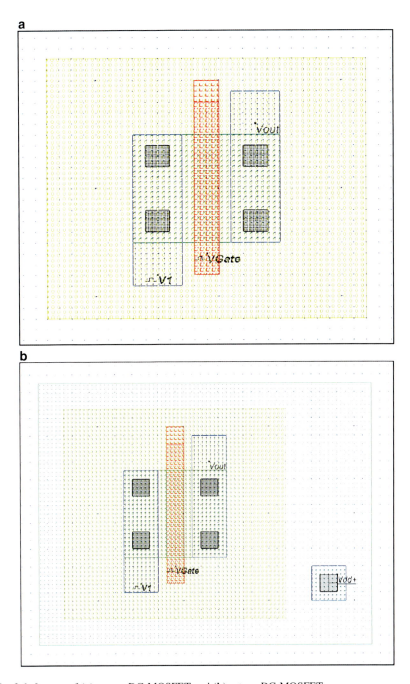

Fig. 3.6 Layout of (**a**) n-type DG MOSFET and (**b**) p-type DG MOSFET

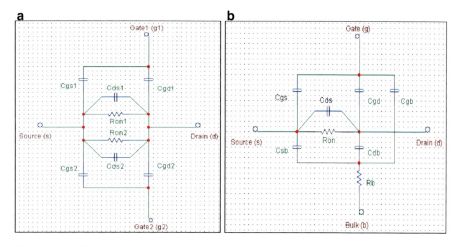

Fig. 3.7 The Circuit Models of (**a**) DG MOSFET and (**b**) SG MOSFET operating as a switch at ON-state

bulk resistance R_b. At the transistors cut-off region C_{ds}, C_{gd}, and C_{gs} increase which direct to lower isolation between the source and drain due to capacitive coupling between these terminals [119]. In Fig. 3.7a, for DG MOSFET, the total capacitance across source to drain is

$$C_{DG} = C_{ds1} + C_{ds2} + \frac{C_{gs1} \cdot C_{gd1}}{C_{gs1} + C_{gd1}} + \frac{C_{gs2} \cdot C_{gd2}}{C_{gs2} + C_{gd2}} \qquad (3.2)$$

and the ON-resistance will be combination of parallel resistances due to gate-1 and gate-2 as follows:

$$R_{DG} = \frac{R_{ON1} \cdot R_{ON2}}{R_{ON1} + R_{ON2}} \qquad (3.3)$$

In Fig. 3.9b, for SG MOSFET, the total capacitance across source to drain is

$$C_{SG} = C_{ds} + \frac{C_{gs} \cdot (C_{gd} + C_{gb})}{C_{gs} + C_{gd} + C_{gb}} + \frac{C_{sb} \cdot C_{db}}{C_{sb} + C_{db}} \qquad (3.4)$$

where C_{gb} is capacitance from gate to bulk connections and the ON-resistance will be only resistance due to single gate:

$$R_{SG} = R_{ON} \qquad (3.5)$$

where

$$R_{ON} = \frac{1}{\mu C_{ox} \frac{W}{L} (V_{gs} - V_{th})} \qquad (3.6)$$

3.5 Resistive and Capacitive Model of DG MOSFET and SG MOSFET

Table 3.1 Comparison of the various circuit parameters of the DG and SG MOSFET for proposed model

Parameters	Double-gate MOSFET	Single-gate MOSFET
Gate/control voltage	1.2 V	1.2 V
Drain to source current ($I_{ds,\ max}$)	80 µA	76 µA
Output voltage stabilization ($V_{out,\ max}$)	0.53 V	0.90 V
Capacitance	C_{DG}	$0.7 C_{DG}$
ON-Resistance (R_{ON})	$0.5 R_{ON}$	R_{ON}
Thickness of oxide layer	3 µm	2 µm
Resistance of poly/gate	68 Ω	32 Ω
No. of capacitors	6	6
Bulk capacitor	No	Yes
Gain (=1 upto)	0.60 V	0.40 V

Since by the calculation of capacitances with (3.2) and (3.4), we obtained that the capacitance $C_{DG} > C_{SG}$, which shows that the isolation is better in double-gate MOSFET compared to that in the single-gate MOSFET. Also, the resistance $R_{DG} < R_{SG}$, which shows that the current flow from source to drain in the double-gate MOSFET is better than single-gate MOSFET. For appropriate working of a switch and to reduce the insertion loss, we can also achieve reduction in ON-resistance with choosing transistor with large transconductance (μ), increasing aspect ratio (W/L), and keeping $V_{gs} - V_{th}$ large as clear from (3.6). Table 3.1 shows the comparison of DG MOSFET and SG MOSFET parameters, which are found near to aspect of radio-frequency switch design with DG MOSFET as compared to the SG MOSFET. Also, the gain is constant up to 0.60 V for DG MOSFET compared to the 0.40 V of SG MOSFET. Here the ON-resistance of DG MOSFET is half of the SG MOSFET, which is useful for DP4T switch.

Gidon [64] has used the MOS transistor model from COMSOL as a template to propose the model of DG MOSFET, for the purpose of resolving the short channel effect problems in MOSFET structures. Such architectures are directly related to the constant reduction of the feature size in microelectronic technology. The drain current vs. gate voltages were considered as parameters.

Gidon [64] has selected these curves in the saturation region at $V_{dd} = 1$ V, so that the transistor will not be in saturation and found the current of 9.80 mA. Tamer and Roy [120] have discussed the CMOS switch structures such as back gate, metal-gate work-function engineering, and gate-isolation processes offer attractive options for the circuit design in multi-gate transistor. With the insertion of extra independent gate processes into fabrication flow allow the exploitation of undoped ultrathin-body-associated strong gate-to-gate coupling in the double-gate MOSFET structures. However under the leakage constraint, the CMOS circuits provide the best power performance trade-off with the symmetric devices, the drain current of 5.0 mA at the gate voltage of 1.0 V. Li and Chou [121] have proposed a unified 2-D density gradient model of a switch, which have simulated sub 10-nm MOSFETs and yield that the MOSFETs with thinner Si-films significantly suppress the short channel effects, but the ON-state current issue suffers. A compromise between the

Table 3.2 Comparison of the drain current for proposed DG MOSFET model with the existing model

References	Drain to source current (mA)
[64]	9.80
[120]	5.00
[121]	3.00
[108]	2.20
[123]	1.75
[124]	1.00
[100]	0.50
DG MOSFET	0.080

Si-film thickness and gate channel length should be maintained at the same time so that an optimal device characteristic could be obtained. Gogineni et al. [122] have presented a comprehensive study of the RF power performance of low power CMOS devices. The device structures with different layouts and widths are studied to understand the effect of the device geometry on RF power performance and show that the power added efficiency (PAE) and output power (P_{out}) decrease with increasing the device width because of the reduction in the maximum frequency (f_{max}). Razavi and Orouji [123] have proposed a model using triple material MOSFET to reduce the short channel effects of nanoscale MOSFET and improving the reliability of the device. However, based on the simulation results, it has been demonstrated that due to the presence of three different materials with different work functions in the gates, the MOSFET switch exhibits reduced short channel effects such as drain-induced barrier lowering and hot carrier effect and improved the reliability. Also, it can be seen that the MOSFET switch leads to simultaneous enhancement of transconductance and reduction of drain conductance which itself leads to higher DC gain in comparison to the conventional DG MOSFET. However, the better reduction of the short channel effects and improvement of the device reliability could be expected by changing the ratio of the gate materials and optimizing them.

Roy [124] has proposed a novel design technique for latch-based voltage mode sense amplifiers using symmetric double-gate devices in sub 50-nm technology. The independent back-gate control of the double-gate device in the pull-down path (other transistors are kept in the connected gate mode) is used to improve the performance and power in sense-amplifier circuits. This proposed design illustrates the fact that selective use of independent control of the front and the back gates in the double-gate devices is very effective in designing an efficient circuit in the nanometer regimes. Hamed et al. [100] have presented the unique and novel examples of low power current mode analog circuit blocks based on DG MOSFET, using mixed mode (device and circuit) TCAD simulations, and illustrate that how the bottom gate of independently driven DG MOSFETs may be used to design and test current mode analog circuits with tunable performance metrics. However, the discussed results are summarized in Table 3.2, which compares the drain current for the control voltage of 1.0 V, with the different existing model, with an intention to maintain required necessary current, resulting in better performance.

3.6 Characteristics of the DG MOSFET with Aspect Ratios

However, the MOSFETs with a lower aspect ratio (*W/L*) are desirable for several reasons. The main reason is to make higher transistors density on a chip area. This results in a chip with the same functionality in a smaller area or more functionality in the same area. Since the fabrication costs for a semiconductor wafer are relatively fixed, the cost per integrated circuits is mainly related to the number of chips that can be produced per wafer. Hence, the smaller ICs allow more chips per wafer and reducing the price per chip. In fact, over the last 30 years the number of transistors per chip has been doubled every 2–3 years once a new technology node is introduced; however, the smaller transistors switches are faster. For example, one approach to size reduction is a scaling of the MOSFET that requires all device dimensions to reduce proportionally. The main device dimensions are the transistor length, width, and the oxide thickness, each (used to) scale with a factor of 0.3 per node. On this way, the transistor channel resistance does not change with scaling, while gate capacitance is cut by a factor of 0.3. Hence, the RC delay of the transistor scales with a factor of 0.3 means it decreases. Independent control of both the gates (front gate and back gate) of the DG MOSFET can be used to improve the performance and to reduce the power loss in the circuits. It can also be used to merge parallel transistors in the noncritical paths, which results the reduction in effective switching capacitance and hence power dissipation [125]. The behavior of a switch depends on the number of controlling gates and additional logic circuit can be implemented into a single transistor. However, independent double-gate transistors can be applied to implement the universal logic functionality within a single transistor. These features make Si-CMOS significant for use in applications that require mixed radio-frequency (RF) and digital systems [126, 127].

For the purpose of 45-nm technology, we have selected two aspect ratios. First, the channel length $L = 0.045$ μm and channel width $W = 22.5$ μm (aspect ratio is 500), and second the channel length $L = 0.045$ μm and channel width $W = 90$ μm (aspect ratio is 2,000). The simulated results for aspect ratio 2,000 are shown in Fig. 3.8a. For the DP4T DG RF CMOS switch design, we consider the control voltage of 1.0 V. At this voltage, the drain current I_{ds} decreases with respect to the bulk voltage. So for higher I_{ds} as of 85 mA, lowest bulk voltage ($V_b = 0$ V) is needed. Also, for the aspect ratio of 500, Fig. 3.9a shows 22 mA of current I_{ds} at control voltage of 1.0 V with lowest bulk voltage ($V_b = 0$ V). Since in the DG MOSFET, bulk voltage is zero, so we can obtain the highest current easily using this proposed switch.

Figures 3.8b and 3.9b, show the threshold voltage for the n-MOSFET of aspect ratio 2,000, with channel length $L = 0.045$ μm and channel width $W = 90$ μm, and for aspect ratio 500, with channel length $L = 0.045$ μm and channel width $W = 22.5$ μm, respectively. However, both the curves represent the same result means this threshold voltage (V_{th}) is 0.3 V at bulk voltage of 0.0 V, whereas for already existing CMOS switch it is 0.7 V. So, the decrement of the threshold

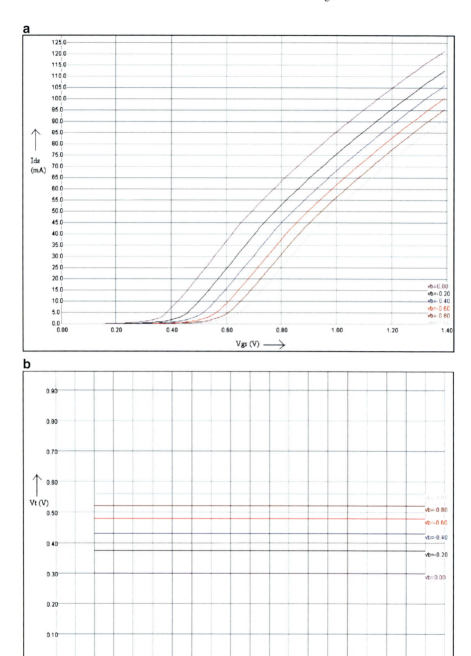

Fig. 3.8 Effect of the aspect ratio (when it is 2000) on the characteristics of DG MOSFET (**a**) drain current with gate to source voltage and (**b**) threshold voltage with the length (nm) of the channel

3.6 Characteristics of the DG MOSFET with Aspect Ratios

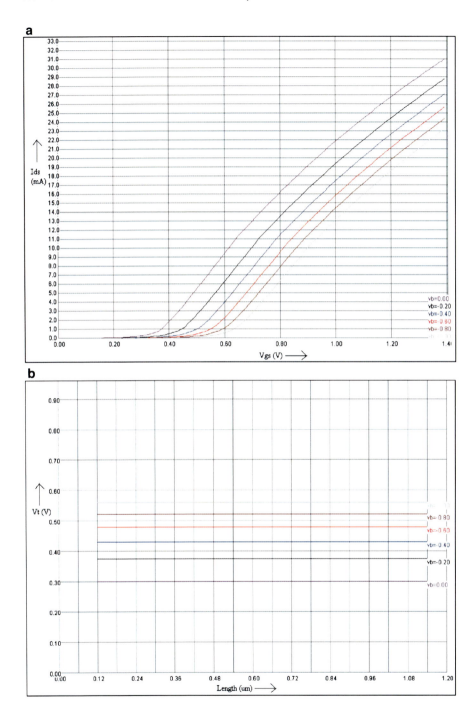

Fig. 3.9 Effect of the aspect ratio (when it is 500) on the characteristics of DG MOSFET (**a**) drain current with gate to source voltage and (**b**) threshold voltage with the length (nm) of the channel

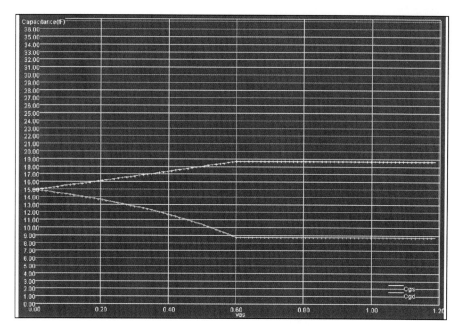

Fig. 3.10 Characteristics of capacitances with drain to source voltage for n-type MOSFET with the aspect ratio 2,000

voltage is a good advantage of 45-nm technology. Since the ultrathin body SOI FETs are capable to achieve better control on the channel by the gate, and hence, reduce the leakage currents and short channel effects.

We can achieve this property using proposed DG MOSFET, because this DG MOSFET has intrinsic or lightly doped body which reduces the threshold voltage variations due to the random dopant fluctuations [128, 129] and enhances the mobility of carriers in the channel region and therefore switch ON-static current. So, we have tried to remove the doping from body or use a pure material. Figure 3.10 shows the gate capacitances with respect to the source and drain. The Gate-source capacitance (C_{gs}) increases with the increase in drain-source voltage (V_{ds}) whereas gate-drain capacitance (C_{gd}) decreases. After the $V_{ds} = 0.60$ V, both capacitances become stable at $C_{gs} = 1.8$ fF and $C_{gd} = 0.8$ fF.

3.7 Design of DG MOSFET with Several Gate-Fingers

The simulations have been performed at 45-nm device structures with the geometries of one gate-finger and ten gate-fingers to understand the effect of device layout and width on RF switch performance using n-type DG MOSFET. The first set of test structures explores the effect of contacting the gate at one end or at both ends of one gate-finger. Figure 3.11a shows the layout of DG MOSFET for one

3.7 Design of DG MOSFET with Several Gate-Fingers

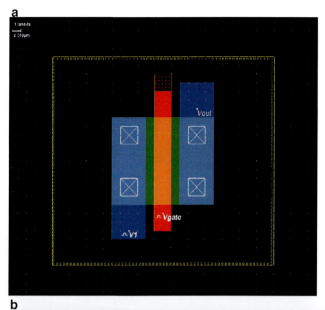

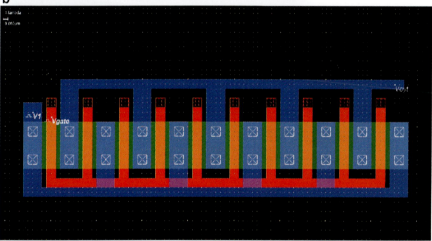

Fig. 3.11 Layout of n-type DG MOSFET for (**a**) $NF = 1$ and (**b**) $NF = 10$

gate-finger ($NF = 1$) and Fig. 3.11b is for the ten gate-fingers ($NF = 10$). Figures 3.12 and 3.13 compare the parameters of the one-gate DG MOSFET structure with that of the ten-gate DG MOSFET structure as a voltage characteristic and drain current characteristics, respectively. The impacts of number of fingers are also clear from Fig. 3.14. For example, one approach to size reduction is a scaling of the MOSFET that requires all device dimensions to reduce proportionally. The scalable CMOS approaches support both n-well and p-well processes. The DG MOSFET is designed for the channel length of 120 nm and the width

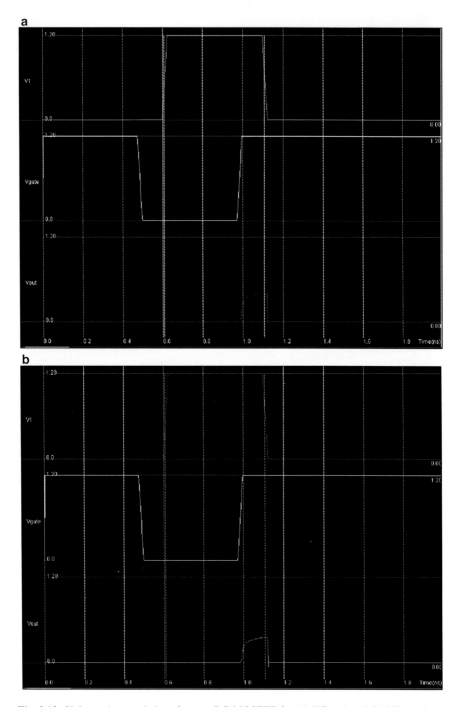

Fig. 3.12 Voltage characteristice of n-type DG MOSFET for (**a**) $NF = 1$ and (**b**) $NF = 10$

3.7 Design of DG MOSFET with Several Gate-Fingers

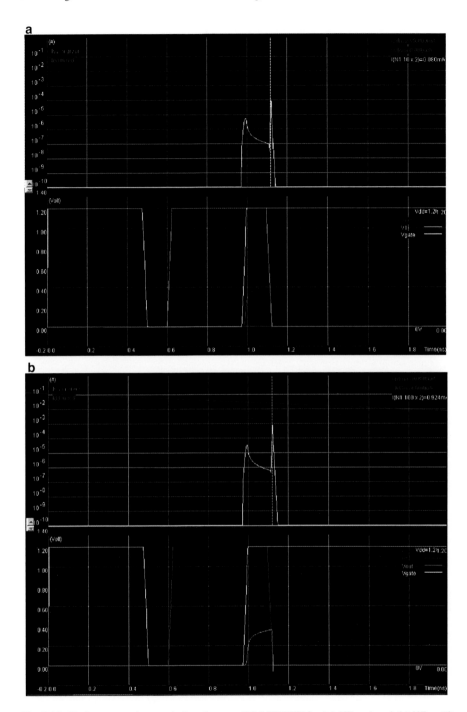

Fig. 3.13 Drain current characteristics of n-type DG MOSFET for (**a**) $NF = 1$ and (**b**) $NF = 10$

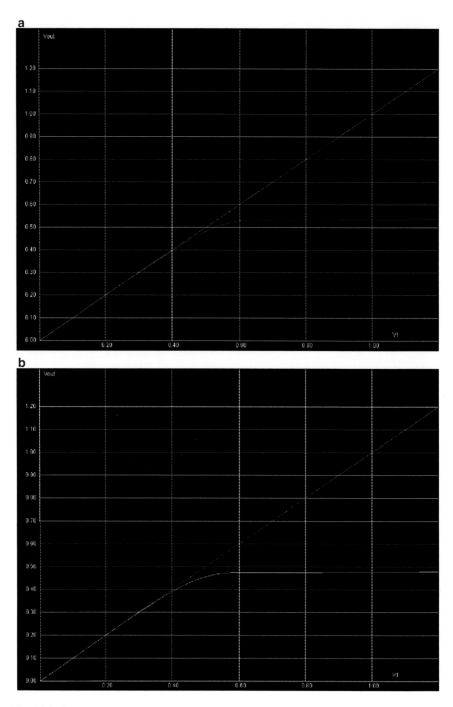

Fig. 3.14 Output voltage characteristics of n-type DG MOSFET for (**a**) $NF = 1$ and (**b**) $NF = 10$

3.7 Design of DG MOSFET with Several Gate-Fingers

Table 3.3 Comparison of the various circuit parameters of the DG MOSFET for $NF = 1$ and $NF = 10$

Parameters	$NF = 1$	$NF = 10$
Gate/control voltage	1.2 V	1.2 V
Output voltage ($V_{out, max}$)	0.42 V	0.36 V
Drain to source current ($I_{ds, max}$)	5 µA	50 µA
Drain to source current ($I_{ds, min}$)	0.10 µA	0.80 µA
V_{out} fixed at V_{in}	0.54 V	0.46 V

of 600 nm [118]. The applied gate voltage of 1.2 V with low level and start time, rise time, fall time, and pulse time for this signal is 0.475 ns, 0.025 ns, 0.025 ns, and 0.475 ns, respectively and V_1 of 1.2 V with high level and start time, rise time, fall time and pulse time for this signal is 0.600 ns, 0.025 ns, 0.025 ns, and 0.475 ns, respectively.

Assuming that DG MOSFET has symmetric-gate structure and voltage applied on both gates are the same. The simulated results are shown in Table 3.3. The device width can be increased by either keeping the number of fingers ($NF = 10$) constant and increasing the unit finger width (WF) from 0.6 to 6.0 µm, or keeping the finger width ($WF = 1.5$ µm) constant and increasing NF from 1 to 10.

The DG MOSFET results in a significantly lower gate resistance but slightly higher gate capacitance, which leads to slightly lower transition frequency (f_T) for the double-gate contact structure compared to the single-gate contact at the same drain current density. However, the maximum frequency (f_{max}) is higher in the double-gate contact structure because the reduction in gate resistance more than that compensates for the increase in the gate capacitance. The improvement in f_{max} is more pronounced at higher current densities. The output voltage stabilizes at 0.54 V while the input voltage is 0.50 V for $NF = 1$ and stabilizes at 0.46 V while the input voltage is 0.39 V for $NF = 10$ which is shown in Fig. 3.14. So, it is better to use higher number of gate-finger in a device.

The presented compact model accounts for the charge quantization within the channel, Fermi statistics, and non-static effects in the transport model. The main finales of this compact switch model are:

a. The DIBL is minimized by the shielding effect of the DG MOSFET, which allows reduction in the channel length from 45 nm.
b. The device transconductance per unit width is maximized by the combination of DG MOSFET and by a strong velocity overshoot, which occurs in response to the abrupt variation of the electric field at the source end of the channel [130].
c. Increase in the device transconductance per unit width can be further strengthened near the drain in view of the short device length.

Consequently, a sustained electron velocity of nearly twice the saturation velocity is obtained. The following observations prove the potential performance of the double-gate device architecture as a switch.

The drain current is described with the idea of Pao and Sah phenomena [131] that includes both the drift and diffusion transport tendencies in the Si-film, resulting in a current description with flat transitions between the linear and saturation operating regions. However, under the approximation that the mobility is independent of the position in the channel, the drain current I_d can be expressed as

$$I_d = \mu \frac{W}{L} \int_0^{V_{ds}} Q_1 dV \tag{3.7}$$

where μ and W are the effective electron mobility and channel width, respectively. L and Q_1 are the effective channel length and total (integrated in the transverse direction) inversion charge density inside the Si-film of a symmetric DG MOSFET at a given location x, which is defined as

$$Q_1 = -2q \int_0^{t_{si}} (n - n_i) dx = -2q \int_{\phi_0}^{\phi_s} \frac{n - n_i}{F} d\phi \tag{3.8}$$

where F is the electric field. Since there is no fixed charge in the undoped body, Q_1 can be taken as being the total charge in the semiconductor [79]:

$$Q_1 = 2\varepsilon_s F_s = -2C(V_{GF} - \phi_s) \tag{3.9}$$

where F_S is the electric field at the surface, and the factor of two comes from the symmetry. An equivalent to the Pao and Sah's equation for the SOI MOSFET may be obtained by substituting (3.9) into (3.8), which yields the following generalized two integral formulations for the drain current:

$$I_d = 2\mu \frac{W}{L} \int_0^{V_{ds}} \int_{\phi_0}^{\phi_s} \frac{qn}{F} d\phi dV \tag{3.10}$$

with $n = n_i e^{\beta(\phi - V)}$. Hence, we can conclude that for the device under test, charge Q_1 in the SG MOSFET is double than that of the DG MOSFET. So the current will be double in the DG MOSFET as compared to the SG MOSFET. The transmission feed line system performance plays an important role in the wireless network coverage. However, the insertion loss measurement is one of the critical measurements used to analyze the transmission feed line installation and performance quality. The insertion loss is given by

$$\left(\frac{R_{ON} + 2Z_0}{2Z_0} \right) \tag{3.11}$$

where Z_0 is the fixed characteristic impedance and taken as 50 Ω. R_{ON} is the resistance of device at ON-state. For DG MOSFET, R_{ON} becomes $R_{ON}/2$ (parallel combination of R_{ON} due to the front gate and back gate). So the insertion loss for the

DP4T DG RF CMOS switch becomes less as compared to SG MOSFET devices. For maximum power transfer the insertion loss should be as small as possible.

With the DP4T DG RF CMOS switch, we have analyzed the output voltage with respect to time as shown in Fig. 3.3. We have presented a comparative analysis of this output voltage for SG MOSFET (Fig. 3.3b) and DG MOSFET (Fig. 3.3a). For SG MOSFET, after closing the switch that is at switch OFF-state condition, the output voltage is 0.74 V at 4.5 ns, whereas for DG MOSFET, the output voltage reduces to zero at 4.5 ns. This fast reduction in output voltage for the DG MOSFET device as DP4T RF CMOS switch is a better switching as compared to that of the SG MOSFET.

A single switch element is characterized for ON/OFF ratio and insertion loss. However, the ON/OFF ratio of a single switch element is $10 \log(2\pi R_{ON} C_{OFF} f_0)$ where R_{ON}, C_{OFF}, and f_0 are ON-static resistance, OFF-static capacitance and the frequency, respectively. Therefore, for the DG MOSFET switch at the high frequency (GHz range) this ratio is higher, means once again switching become fast for DP4T DG RF CMOS switch.

3.8 Model of Series and Parallel Combination for Double-Gate MOSFET

In general, the DG MOSFET devices are employed with tied gate controlled (TGC) or independently gate controlled (IGC). The tied gate controlled circuit topology resembles conventional planar CMOS configuration and provides higher current density with potential drive capability as well as more compact layout area. For the independent gate-controlled MOSFET, the symmetrical double-gate devices can reduce the transistor count and result the significant reduction in the chip area to implement a given logic function [132]. However, independent control of both gates of the DG MOSFET can be used to improve the performance and reduces the power loss in the circuits. It can also be used to merge parallel transistors in the noncritical paths. This results the reduction in an effective switching capacitance and hence power dissipation [133]. The behavior of a switch depends on the number of controlling gates and additional logic circuit can be implemented into a single transistor. The independent double-gate transistors can be applied to implement universal logic functionality within a single transistor. These features make Si-CMOS more significant for the use in applications that require mixed RF and digital systems [134–136].

However, independently controlled gate transistors reduces the number of transistors from circuit design perception, as two transistors connected in series or parallel combination can be compounded into the one transistor as shown in Fig. 3.15. This procedure reduces the transistors stack height as well as chip area and power consumption. There is a trade-off between area reduction and gate delay and leakage has to be selected, when using transistors with independently controlled gates. Table 3.4 shows the transistor status for IGC and TGC with logic design.

Fig. 3.15 Conversion of the series and parallel combination of n-MOSFET/p-MOSFET to DG MOSFET. Case 1. Series combination of n-MOSFET to DG MOSFET. Case 2. Series combination of p-MOSFET to DG MOSFET. Case 3. Parallel combination of n-MOSFET to DG MOSFET. Case 4. Parallel combination of p-MOSFET to DG MOSFET

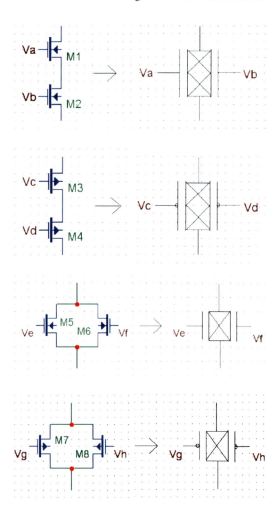

Table 3.4 Design for independent gate configuration (IGC) and tied gate configuration (TGC)

| Figure 3.15 | Type | Logic | Independent gate configuration ||| Tied gate configuration |||
			Gate 1	Gate 2	Transistor status	Gate 1	Gate 2	Transistor status
Case 1	N	AND	Low	High	OFF	High (Va)	High (Vb)	ON
Case 2	P	AND	High	Low	OFF	High (Vc)	High (Vd)	OFF
Case 3	N	OR	Low	High	ON	Low (Ve)	Low (Vf)	OFF
Case 4	P	OR	High	Low	ON	Low (Vg)	Low (Vh)	ON

3.9 Conclusions

Table 3.5 An effective aspect ratio for different combination of transistors as shown in Fig. 3.15

Figure 3.15	Effective aspect ratio (W/L)
Case 1	0.5
Case 2	2.0
Case 3	0.5
Case 4	2.0

The use of IGC is better for gates with small fan out and short capacitive wire loads. The circuit of Fig. 3.15 uses independently controlled gates which are investigated by simulation in [137]. With this double-gate MOSFET, we can design all the connection of Fig. 3.15 with the aspect ratio (W/L) as shown in Table 3.5.

3.9 Conclusions

In this chapter, a symmetrical DG MOSFET has been modeled. For the purpose of RF/microwave switch, we have discussed the process to minimize the control voltage, capacitances for isolation, and the resistance for the switching condition. From the discussions in preceding sections we found a better DG MOSFET as compared to the SG MOSFET. It concludes that capacitance of a MOSFET is changing with the change of a bias voltage. Moreover, it depends on the aspect ratio, as greater the aspect ratio greater its capacitance. The capacitance changes of two gate capacitance in same ratio as in SG MOSFET which is also shown in the graphs.

The values of capacitances at different various gate voltages vary just like in SG MOSFET. For increasing the gate voltage, the drain current increases, hence the contact resistance decreases which increases the cut-off frequency. Therefore, for the purpose of RF switch, where control voltage should be low and then current flow will be less and in terms of contact resistance, it will increase with increase in number of gate-fingers. So, in the application of RF switch, we have tried to increase the gate-finger. Since the operating frequencies of the RF switches are in the order of GHz, it is very useful for the modern wireless communication systems [80].

From the simulated result of DG MOSFET for one gate-finger and for ten gate-fingers, highest drain current can be easily achieved by using higher gate-fingers. At higher technology, the ultrathin body SOI FETs employ very thin Silicon body to achieve better control of the channel by the gate and hence, reduced the leakage and short channel effects. With the use of intrinsic or lightly doped body, in the DG

MOSFET, the reduction in the threshold voltage occurs due to the random dopant fluctuations. As at higher gate-fingers, it also enhances the mobility of the carriers in the channel region as compared to the lower gate-fingers and therefore increment in drain current occurs at higher NF. Therefore, we can get a better result by using higher gate-fingers for the DG MOSFET at 45-nm technology, as it has intrinsic or lightly doped body for the application of DP4T DG RF CMOS switch.

References

1. Y. Taur and T. H. Ning, *Fundamentals of Modern VLSI Devices*, New York: Cambridge Univ. Press, 1998.
2. J. Jacobs and D. Antoniadis, "Channel profile engineering for MOSFET's with 100 nm channel lengths," *IEEE Trans. on Electron Devices*, vol. 42, no. 5, pp. 870–875, May 1995.
3. 2004 International Technology Roadmap for Semiconductors (ITRS), www.public.itrs.net
4. K. Takeuchi, "A study of the threshold voltage variation for ultra-small bulk and SOI CMOS," *IEEE Trans. Electron Devices*, vol. 48, no. 9, pp. 1995–2001, Sept. 2001.
5. H. Wong, "Device design considerations for double-gate, ground-plane, and single-gated ultra-thin SOI MOSFET's at the 25 nm channel length generation," *Proc. of Int. Electron Devices Meeting*, San Francisco, CA, USA, 6–9 Dec. 1998, pp. 407–410.
6. L. Chang, Choi Yang, Ha Daewon, P. Ramade, Xiong Shiying, J. Bokor, Hu Chenming, and King Tsu, "Extremely scaled silicon nano CMOS devices," *Proc. of IEEE*, vol. 91, no. 11, pp. 1860–1873, Nov. 2003.
7. H. S. P. Wong, D. J. Frank, P. M. Solomon, "Device design considerations for double-gate, ground-plane, single-gated ultra-thin SOI MOSFET at the 25nm channel length generation," *Proc. of Int. Electron Devices Meeting*, San Francisco, CA, USA, 6–9 Dec. 1998, pp. 407–410.
8. E. Nowak, I. Aller, T. Ludwig, Kim Keunwoo, K. Bernstein, "Turning silicon on its edge," *IEEE Circuits and Device Magazine*, vol. 20, no. 1, Jan/Feb 2004, pp. 20–31.
9. S. Tang, L. Chang, N. Lindert, H. Xuejue, J. Bokor, H. Chenming, "FinFET-a quasi-planar double-gate MOSFET," *Proc. of IEEE Int. Solid-State Circuits Conference*, San Francisco, CA, USA, 5–7 Feb. 2001, pp. 118–119.
10. D. M. Fried, E. J. Nowak, J. Kedzierski, J. S. Duster, and K. T. Komegay, "A Fin-type independent-double-gate NFET," *Proc. of Device Research Conference*, Salt Lake City, UT, USA, 23–25 June 2003, pp. 45–46.
11. L. Mathew, Y. Du, A. Thean, M. Sadd, A. Vandooren, C. Parker, T. Stephens, R. Mora, J. Hughes, R. Shimer, S. Jallepalli, W. Zhang, J. G. Fossum, B. Y. Nguyen, and J. Mogab, "CMOS vertical multiple independent gate field effect transistors (MIGFET)", *Proc. of Int. SOI Conf.*, South Carolina, USA, 4–7 Oct. 2004, pp. 187–189.
12. S. Mukhopadhyay, H. Mahmoodi, and K. Roy, "Design of high performance sense amplifier using independent gate control in sub-50 nm double-gate MOSFET," *Proc. of 6th Int. Symp. on Quality of Electronic Design*, San Jose, CA, USA, 21–23 March 2005, pp. 490–495.
13. S. Mukhopadhyay, H. Mahmoodi, and K. Roy, "High performance and low power domino logic using independent gate control in double-gate SOI MOSFETs," *Proc. of IEEE Int. SOI Conf.*, South carolina, USA, 4–7 Oct. 2004, pp. 67–68.
14. S. Mukhopadhyay, H. Mahmoodi, and K. Roy, "Independent gate skewed logic in double-gate SOI technology," *Proc. of IEEE Int. SOI Conference*, Honolulu, Hawaii, USA, 3–6 Oct. 2005, pp. 83–84.

References

15. T. Cakici and K. Roy, "A low power four transistor Schmitt trigger for asymmetric double-gate fully depleted SOI devices," *Proc. of IEEE Int. SOI Conference*, CA, USA, 29 Sept.–2 Oct. 2003, pp. 21–22.
16. Bo Yu, Jooyoung Song, Yu Yuan, and Yuan Taur, "A unified analytic drain current model for multiple-gate MOSFETs," *IEEE Trans. on Electron Devices*, vol. 55, no. 8, pp. 2157–2163, Aug. 2008.
17. Tae H. Kim, J. Keane, Hanyong Eom, and C. H. Kim, "Utilizing reverse short-channel effect for optimal subthreshold circuit design," *IEEE Trans. on Very Large Scale Integration Systems*, vol. 15, no. 7, pp. 821–829, July 2007.
18. O. Moldovan, D. Jimenez, J. Guitart, F. A. Chaves, and B. Iniguez, "Explicit analytical charge and capacitance models of undoped double-gate MOSFETs," *IEEE Trans. on Electron Devices*, vol. 54, no. 7, pp. 1718–1724, July 2007.
19. T. Park, S. Choi, D. H. Lee, J. R. Yoo, B. C. Lee, J. Y. Kim, C. G. Lee, S. H. Hong, S. J. Hynn, Y. G. Shin, J. N. Han, I. S. Park, U. I. Chung, J. T. Moon, E. Yoon, and J. H. Lee, "Fabrication of body-tied FinFETS (Omega MOSFETs) using bulk Si wafers," *Proc. of Symp. on VLSI Technology*, Honolulu, Hawaii, USA, 11–13 June 2002, pp. 135–136.
20. S. Monfray and T. Skotnicki, "50 nm gate all around (GAA) silicon on nothing (SON) devices: A simple way to co-integration of GAA transistors with bulk MOSFET process," *Proc. of Symp. on VLSI Technology*, Honolulu, USA, 11–13 June 2002, pp. 108–109.
21. Antonioramon Lazaro and Benjamin Iniguez, "RF and noise model of gate-all-around MOSFETs," *Semiconductor Science and Technology*, vol. 23, no. 7, pp. 1-12, July 2008.
22. Aditya Bansal and Kaushik Roy, "Modeling and optimization of fringe capacitance of nanoscale DGMOS devices," *IEEE Trans. on Electron Device*, vol. 52, no. 2, pp. 256–262, Feb. 2005.
23. Antonioramon Lazaro and Benjamin Iniguez, "High-frequency compact analytical noise model of gate-all-around MOSFETs," *Semiconductor Science and Technology*, vol. 25, no. 3, pp. 1–10, March 2010.
24. T. Skotnicki, J. A. Hutchby, T. King, H. Wong, and F. Boeuf, "The end of CMOS scaling: toward the introduction of new materials and structural changes to improve MOSFET performance," *IEEE Circuits and Devices Magazine*, vol. 21, no. 1, pp. 16–26, Jan. 2005.
25. H. Borli, K. Vinkenes, and T. A. Fjeldly, "Physics based capacitance modeling of short channel double-gate MOSFETs," *Physica Status Solidi C*, vol. 5, no. 12, pp. 3643–3646, Dec. 2008.
26. F. Lime, B. Iniguez, and O. Moldovan, "A quasi-two-dimensional compact drain-current model for undoped symmetric double-gate MOSFETs including short-channel effects," *IEEE Trans. on Electron Devices*, vol. 55, no. 6, pp. 1441–1448, June 2008.
27. Meishoku Masahara, "Fabrication and characterization of vertical-type, self-aligned asymmetric double-gate MOSFET," *Applied Physics Letters*, vol. 86, no. 12, pp. 512–513, Dec. 2005.
28. International Technology Roadmap for Semiconductor 2006 edition (http://public.itrs.net)
29. P. M. Solomon, "Two gates are better than one," *IEEE Circuits Devices Mag.*, vol. 19, no. 1, pp. 48–62, Jan. 2003.
30. A. Kranti, T. M. Chung, D. Flandre, and J. P. Raskin, "Analysis of quasi double gate method for performance prediction of deep submicron double gate SOI MOSFETs," *Semicond. Sci. Technol.*, vol. 20, no. 5, pp. 423–429, May 2005.
31. J. G. Fossum, L. Ge, and M. H. Chiang, "Speed superiority of scaled double-gate CMOS," *IEEE Trans. Electron Devices*, vol. 49, no. 5, pp. 808–811, May 2002.
32. H. S. P. Wong, D. J. Frank, Y. Taur and J. M. C. Stork, "Design and performance considerations for sub-0.1 μm double-gate SOI MOSFETs," *Proc. of IEEE Int. Electron Devices Meeting*, San Francisco, CA, USA, 11–14 Dec. 1994, pp. 747–750.
33. F. Balestra, S. Cristoloveanu, M. Benachir, J. Birni and T. Elewa, "Double-gate silicon-on-insulator transistor with volume inversion: a new device with greatly enhanced performance," *IEEE Electron. Device Lett.*, vol. 8, no. 9, pp. 410–412, Sept. 1987.

34. A. Kranti, T M Chung and J P Raskin, "Analysis of static and dynamic performance of short channel double gate SOI MOSFETs for improved cut-off frequency," *Japan. J. Appl. Phys.*, vol. 44, pp. 2340–2346, 2005.
35. A. Kranti, T M Chung and J P Raskin, "Double gate SOI MOSFET, considerations for improved cut-off frequency," *Proc. of Int. Conf. on Solid State Devices and Materials*, Tokyo, Japan, 14–17 Sept. 2004, pp. 784–785.
36. Viranjay M. Srivastava, K. S. Yadav, and G. Singh, "Design and performance analysis of double-gate MOSFET over single-gate MOSFET for RF switch," *Microelectronics Journal*, vol. 42, no. 3, pp. 527–534, March 2011.
37. J. M. Hergenrother, D. Monroe, F. P. Klemens, A. Komblit, G. R. Weber, W. M. Mansfield, M. R. Baker, F. H. Baumann, K. J. Bolan, J. E. Bower, N. A. Ciampa, R. A. Cirelli, J. I. Colonell, D. J. Eaglesham, J. Frackoviak, H. J. Gossmann, M. L. Green, S. J. Hillenius, C. A. King, R. N. Kleiman, W. C. Lai, T. C. Lee, R. C. Liu, H. L. Maynard, M. D. Morris, C. S. Pai, C. S. Rafferty, J. M. Rosamilia, T. W. Sorsch, and H. H. Vuong, "The vertical replacement gate (VRG) MOSFET: a 50-nm vertical MOSFET with lithography independent gate length," *Proc. of IEEE Int. Electron Devices Meeting*, Washington, DC, USA, 5–8 Dec. 1999, pp. 75–78.
38. B. Yu, H. Wang, A. Joshi, Q. Xiang, E. Ibok, and M. R. Lin, "15 nm gate length planar CMOS transistor," *Proc. of IEEE Int. Electron Devices Meeting*, San Francisco, CA, USA, 2001, pp. 937–940.
39. J. Kedzierski, P. Xuan, E. H. Anderson, J. Bokor, T. K. King, and C. Hu, "Complementary silicide source/drain thin body MOSFETs for the 20 nm gate length regime," *Proc. of IEEE Int. Electron Devices Meeting*, San Francisco, CA, USA, 2001, pp. 57–60.
40. D. Hisamoto, T. Kaga, Y. Kawamoto, and E. Takeda, "A fully depleted lean-channel transistor (DELTA)—a novel vertical ultra thin SOI MOSFET," *Proc. of IEEE Int. Electron Devices Meeting*, San Francisco, CA, USA, 1989, pp. 833–836.
41. J. Kedzierski, "High-performance symmetric-gate and CMOS-compatible Vt asymmetric-gate FinFET devices," *Proc. of IEEE Int. Electron Devices Meeting*, San Francisco, CA, USA, 2001, pp. 437–440.
42. J. Kedzierski, M. Ieong, E. Nowak, and H. S. P. Wong, "Extension and source/drain design for high performance FinFET devices," *IEEE Trans. on Electron Devices*, vol. 50, no. 4, pp. 952–958, April 2001.
43. A. Dixit, A. Kottantharayil, N. Collaert, and De Meyer, "Analysis of the parasitic S/D resistance in multiple-gate FETs," *IEEE Trans. on Electron Devices*, vol. 52, no. 6, pp. 1132–1140, June 2005.
44. F. Boeuf, "16 nm planar NMOSFET manufacturable within state of the art CMOS process thanks to specific design and optimization," *Proc. of IEEE Int. Electron Devices Meeting*, San Francisco, CA, USA, 2–5 Dec. 2001, pp. 29.5.1–29.5.4.
45. A. Asenov, "Random dopant induced threshold voltage lowering and fluctuations in sub-0.1 μm MOSFET's: a 3D 'atomistic' simulation study," *IEEE Trans. on Electron Devices*, vol. 45, no. 12, pp. 2505–2513, Dec. 1996.
46. H. Liu, Z. Xiong, and J. Sin, "Implementation and characterization of the double-gate MOSFET using lateral solid-phase epitaxy," *IEEE Trans. on Electron Devices*, vol. 50, no. 6, pp. 1552–1555, June 2003.
47. M. Masahara, "Ultrathin channel vertical DG MOSFET fabricated by using ion bombardment retarded etching," *IEEE Trans. on Electron Devices*, vol. 51, no. 12, pp. 2078–2085, Dec. 2004.
48. G. Tsutsui, M. Saitoh, T. Saraya, T. Nagumo and T. Hiramoto, "Mobility enhancement due to volume inversion in (110)-oriented ultra-thin body double-gate nMOSFETs with body thickness less than 5 nm," *Proc. of IEEE Int. Electron Devices Meeting*, San Francisco, CA, USA, 5–7 Dec. 2005, pp. 729–732.
49. J. G. Fossum, "Physical insights on nanoscale multi-gate CMOS design," *Solid State Electron.*, vol. 51, no. 2, pp. 188–194, Feb. 2007.

References

50. A. Kranti, T. Chung, D. Flandre, and J. P. Raskin, "Laterally asymmetric channel engineering in fully depleted double gate SOI MOSFETs for high performance analog applications," *Solid State Electron.*, vol. 48, no. 6, pp. 947–959, June 2004.
51. M. Chiang, K. Kim, C. Tretz, and C. T. Chuang, "Novel high-density low-power logic circuit techniques using DG devices," *IEEE Trans on Electron Devices*, vol. 52, no. 10, pp. 2339–2342, Oct. 2004.
52. L. Wei, R. Zhang, K. Roy, Z. Chen, and D. B. Janes, "Vertically integrated SOI circuits for low-power and high-performance applications," *IEEE Trans. on Very Large Scale Integration Systems*, vol. 10, no. 3, pp. 351–362, March 2002.
53. S. Xiong and J. Bokor, "Sensitivity of double-gate and FinFET Devices to process variations," *IEEE Trans on Electron Devices*, vol. 50, no. 11, pp. 2255–2261, Nov. 2001.
54. J. Widiez, J. Lolivier, M. Vinet, T. Poiroux, B. Previtali, F. Dauge, M. Mouis, and S. Deleonibus, "Experimental evaluation of gate architecture influence on DG SOI MOSFETs performance," *IEEE Trans. on Electron Devices*, vol. 52, no. 8, pp. 1772–1779, Aug. 2001.
55. T. Sekigawa and Y. Hayashi, "Calculated threshold-voltage characteristics of an XMOS transistor having an additional bottom gate," *Solid State Electronics*, vol. 27, no. 8–9, pp. 827–828, Aug.–Sept. 1984.
56. O. Moldovan, A. Cerdeira, D. Jimenez, J. P. Raskin, V. Kilchytska, D. Flandre, N. Collaert and B. Iniguez, "Compact model for highly-doped double-gate SOI MOSFETs targeting baseband analog applications," *Solid State Electronics*, vol. 51, no. 5, pp. 655–661, May 2007.
57. A. Kawamoto, S. Sato and Y. Omura, "Engineering S/D diffusion for sub-100-nm channel SOI MOSFETs," *IEEE Trans. on Electron Devices*, vol. 51, no. 6, pp. 907–913, June 2004.
58. Viranjay M. Srivastava, K. S. Yadav, and G. Singh, "Analysis of double gate CMOS for double-pole four-throw RF switch design at 45-nm technology," *J. of Computational Electronics*, vol. 10, no. 1–2, pp. 229–240, June 2011.
59. T. C. Lim and G. A. Armstrong, "Scaling issues for analogue circuits using double gate SOI transistors," *Solid State Electronics*, vol. 51, no. 2, pp. 320–327, Feb. 2007.
60. A. Kranti and G. A. Armstrong, "Optimization of source/drain doping extension region doping profile for the suppression of short channel effects in sub-50 nm double gate MOSFETs with high-κ gate dielectrics," *Semiconductor Science and Technology*, vol. 21, no. 12, pp. 1563–1572, 2006.
61. A. Kranti and G. A. Armstrong, "Performance assessment of nanoscale double and triple gate FinFETs," *Semiconductor Science and Technology*, vol. 21, no. 4, pp. 409–421, 2006.
62. A. Kranti and G. A. Armstrong, "Design and optimization of FinFETs for ultra low-voltage analog applications," *IEEE Trans. on Electron Devices*, vol. 54, no. 12, pp. 3308–3316, Dec. 2007.
63. Viranjay M. Srivastava, K. S. Yadav, and G. Singh, "Double-pole four-throw CMOS switch design with double-gate transistor," *Proc. of 2010 Annual IEEE India Conf.*, India, 17–19 Dec. 2010, pp. 1–4.
64. Serge Gidon, "Double-gate MOSFET modeling," *Proc. of the COMSOL Multiphysics User's Conf.*, Paris, 2005, pp. 1–4.
65. Huaxin Lu and Yuan Taur, "An analytic potential model for symmetric and asymmetric DG MOSFETs," *IEEE Trans. on Electron Devices*, vol. 53, no. 5, pp. 1161–1168, May 2006.
66. Yuan Taur, "Analytic solutions of charge and capacitance in symmetric and asymmetric double-gate MOSFETs," *IEEE Trans. on Electron Devices*, vol. 48, no. 12, pp. 2861–2869, Dec. 2001.
67. Keunwoo Kim and Jerry G. Fossum, "Optimal double-gate MOSFETs: symmetrical or asymmetrical gates," *Proc. of Int. Conf. on SOI*, California, USA, 4-7 Oct. 1999, pp. 98–99.
68. Keunwoo Kim and Jerry G. Fossum, "Double-gate CMOS: symmetrical versus asymmetrical gate devices," *IEEE Trans. on Electron Devices*, vol. 48, no. 2, pp. 294–299, Feb. 2001.

69. Viktor Sverdlov, "Scaling, power consumption, and mobility enhancement techniques," Edited by S. Selberherr, *Strain induced effects in advanced MOSFET*, Springer-Verlag Wien, New York, pp. 5–22, 2011.
70. Li Qiang and Y. P. Zhang, "CMOS T/R switch design: towards ultra wideband and higher frequency," *IEEE J. of Solid State Circuits*, vol. 42, no. 3, pp. 563–570, March 2007.
71. Stephen A. Maas, *The RF and Microwave Circuit Design Cookbook*, 1st Edition, Artech Publication, London, 1998.
72. Huaxin Lu, Wei Yuan Lu, and Yuan Taur, "Effect of body doping on double-gate MOSFET characteristics," *Semiconductor Science and Technology*, vol. 23, no. 1, pp. 1–6, Jan. 2008.
73. Amalendu Bhushan Bhattacharyya, *Compact MOSFET Models for VLSI Design*, 1st Edition, Wiley, Singapore, 2009.
74. K. Takeuchi, R. Koh, and T. Mogami, "A study of the threshold voltage variation for ultra small bulk and SOI CMOS," *IEEE Trans. on Electron Devices*, vol. 48, no. 9, pp. 1995–2001, Sept. 2001.
75. Hamid Mahmoodi, Saibal Mukhopadhyay, Hari Ananthan, Aditya Bansal, Tamer Cakici, and Kaushik Roy, "Double-gate SOI devices for low power and high performance applications," *Proc. of IEEE Int. Conf. on Computer Aided Design*, San Jose, California, USA, 10 Nov. 2005, pp. 217–224.
76. Hamid Mahmoodi, Saibal Mukhopadhyay, and Kaushik Roy, "Double-gate SOI devices for low power and high performance applications," *Proc. of 19th Int. Conf. on VLSI Design*, India, 3–7 Jan. 2006, pp. 445–452.
77. L. Chang, J. Boko, and T. King, "Extremely scaled silicon nano CMOS devices," *Proc. of IEEE*, vol. 91, no. 11, pp. 1860–1873, Nov. 2003.
78. B. Iniguez, T. A. Fjeldly, A. Lazaro, F. Danneville, and M. J. Deen, "Compact modeling solutions for nanoscale double-gate and gate-all-around MOSFETs," *IEEE Trans. on Electron Devices*, vol. 53, no. 9, pp. 2128–2142, Sept. 2006.
79. Hamid Mahmoodi, Saibal Mukhopadhyay, and Kaushik Roy, "A novel high performance and robust sense amplifier using independent gate control in sub 50-nm double-gate MOSFET," *IEEE Trans. on Very Large Scale Integration Systems*, vol. 14, no. 2, pp. 183–192, Feb. 2006.
80. Hamid Mahmoodi, Saibal Mukhopadhyay, and Kaushik Roy, "High performance and low power domino logic using independent gate control in double-gate SOI MOSFETs," *Proc. of IEEE Int. Conf. on SOI*, South Carolina, USA, 4–7 Oct. 2004, pp. 67–68.
81. Hamid Mahmoodi, Saibal Mukhopadhyay, T. Cakici, and Kaushik Roy "Independent gate skewed logic in double-gate SOI technology," *Proc. of IEEE Int. Conf. on SOI*, Honolulu, Hawaii, USA, 6 Oct. 2005, pp. 83–84.
82. A. delmo O. Conde and Francisco J. Sanchez, "Generic complex variable potential equation for the undoped asymmetric independent double-gate MOSFET," *Solid State Electronics*, vol. 57, no. 1, pp. 43–51, March 2011.
83. Huaxin Lu, Wei Wang, Jooyoung Song, Shih Hsien Lo, and Yuan Taur, "Compact modeling of quantum effects in symmetric double-gate MOSFETs," *Microelectronics Journal*, vol. 41, no. 10, pp. 688–692, Oct. 2010.
84. V. P. Trivedi, J. G. Fossum, and W. Zhang, "Threshold voltage and bulk inversion effects in nonclassical CMOS devices with undoped ultra-thin bodies," *Solid State Electronics*, vol. 51, no. 1, pp. 170–178, Jan. 2007.
85. B. Doyle, B. Boyanov, S. Datta, M. Doczy, S. Hareland, B. Jin, J. Kavalieros, T. Linton, R. Rios, and R. Chau, "Tri-Gate Fully-Depleted CMOS Transistor: Fabrication, Design, Layout," *Symp. on VLSI Tech. Dig.*, Kyoto, Japan, 10–12 June 2003, pp. 133–134.
86. Y. Taur and T. H. Ning, *Fundamentals of Modern VLSI Devices*. New York: Cambridge University Press, 1998.
87. A. Tsormpatzoglou, D. H. Tassis, G. Pananakakis, and N. Collaert, "Analytical modeling for the current voltage characteristics of undoped or lightly doped symmetric double-gate MOSFETs," *Microelectronic Engineering*, vol. 87, no. 9, pp. 1764–1768, Nov. 2010.

References

88. Oana Moldovan, Benjamin Iniguez, A. Cerdeira, and M. Estrada, "Modeling of potentials and threshold voltage for symmetric doped double-gate MOSFETs," *Solid State Electronics*, vol. 52, no. 5, pp. 830–837, May 2008.
89. J. Kedzierski, D. M. Fried, E. J. Nowak, and T. Kanarsky, "High performance symmetric gate and CMOS compatible asymmetric gate FinFET device," *Proc. of Int. Electron Devices Meeting*, Washington, 2–5 Dec. 2001, pp. 437–440.
90. S. Balasubramanian, J. L. Garrett, V. Vidya, and T. J. King, "Energy-delay optimization of thin body MOSFETs for the sub 15-nm regime," *Proc. of IEEE Int. Conf. SOI*, South Carolina, 4–7 Oct. 2004, pp. 27–29.
91. Thomas H. Lee, *The Design of CMOS Radio-Frequency Integrated Circuits*, 2nd Edition, Cambridge University Press, USA, 2004.
92. Keunwoo Kim, K. K. Das, and Ching Chuang, "Nanoscale CMOS circuit leakage power reduction by double-gate devices," *Proc. of Int. Symp. on Low Power Electronics and Design*, California, USA, 9–11 Aug. 2004, pp. 102–107.
93. Leland Chang, K. J. Yang, and Chenming Hu, "Direct tunneling gate leakage current in double-gate and ultrathin body MOSFETs," *IEEE Trans. on Electron Devices*, vol. 49, no. 12, pp. 2288–2295, Dec. 2002.
94. Keunwoo Kim, E. Nowak, T. Ludwig, and Ching Chuang, "Turning Silicon on its edge: double-gate MOSFET and FinFET," *IEEE Circuits and Device Magazine*, vol. 20, no. 1, pp. 20–31, Feb. 2004.
95. Keunwoo Kim, Saibal Mukhopadhyay, Ching Chuang, and Kaushik Roy, "Modeling and analysis of leakage currents in double-gate technologies," *IEEE Trans. on Computer Aided Design of Integrated Circuits and Systems*, vol. 25, no. 10, pp. 2052–2061, Oct. 2006.
96. U. Monga and T. A. Fjeldly, "Compact subthreshold current modeling of short-channel nanoscale double-gate MOSFET," *IEEE Trans. on Electron Devices*, vol. 56, no. 7, pp. 1533–1537, July 2009.
97. F. Schwierz and C. Schippel, "Performance trends of Si-based RF transistors," *Microelectronics Reliability*, vol. 47, no. 2–3, pp. 384–390, Feb.-March 2007.
98. Hari Ananthan, Aditya Bansal, and Kaushik Roy, "Analysis of drain to body band to band tunneling in double-gate MOSFET," *Proc. of IEEE Int. Conf. on SOI*, Honolulu, Hawaii, USA, 3–6 Oct. 2005, pp. 159–160.
99. Oana Moldovan, Antonio Cerdeira, J. P. Raskin, and Benjamin Iniguez, "Compact model for highly doped double-gate SOI MOSFETs targeting baseband analog applications," *Solid State Electronics*, vol. 51, no. 5, pp. 655–661, May 2007.
100. Hesham Hamed, Savas Kaya, Janusz, and A. Starzyk, "Use of nano-scale double-gate MOSFETs in low-power tunable current mode analog circuits," *J. of Analog Integrated Circuits, Signal Processing*, vol. 54, no. 3, pp. 211–217, March 2008.
101. F. Balestra and T. Elewa, "Double-gate silicon on insulator transistor with volume inversion: A new device with greatly enhanced performance," *IEEE Electron Device Letter*, vol. 8, no. 9, pp. 410–412, Sept. 1987.
102. J. P. Denton and G.W. Neudeck, "Fully depleted dual-gate thin film SOI p-MOSFET's fabricated in SOI islands with an isolated buried polysilicon back-gate," *IEEE Electron Device Letter*, vol. 17, no. 11, pp. 509–511, Nov. 1996.
103. Y. Naveh and K. K. Likharev, "Modeling of 10 nm scale ballistic MOSFETs," *IEEE Electron Device Letter*, vol. 21, no. 5, pp. 242–244, May 2000.
104. A. Rahman and M. S. Lundstrom, "A compact scattering model for the nanoscale double-gate MOSFET," *IEEE Trans. on Electron Devices*, vol. 49, no. 3, pp. 481–489, March 2002.
105. L. Ge and Jerry G. Fossum, "Analytical modeling of quantization and volume inversion in thin Si film DG MOSFETs," *IEEE Trans. on Electron Devices*, vol. 49, no. 2, pp. 287–294, Feb. 2002.
106. Y. Omura and K. Kishi, "Quantum-mechanical effects on the threshold voltage of ultra thin-SOI nMOSFETs," *IEEE Electron Device Letter*, vol. 14, no. 12, pp. 569–571, Dec. 1993.

107. Benjamin Iniguez, David Jimenez, Jaume Roig, Hamdy A. Hamid, and Josep Pallares, "Explicit continuous model for long channel undoped surrounding gate MOSFETs," *IEEE Trans. on Electron Devices*, vol. 52, no. 8, pp. 1868–1873, Aug. 2005.
108. Y. Ye and Y. Cao, "Random variability modeling and its impact on scaled CMOS circuits," *J. of Computational Electronics*, vol. 9, no. 3–4, pp. 108–113, Dec. 2010.
109. P. Dollfus, and Retailleau, "Thermal noise in nanometric DG MOSFET," *J. of Computational Electronics*, vol. 5, no. 4, pp. 479–482, Dec. 2006.
110. P. Dollfus, "Sensitivity of single and double-gate MOS architectures to residual discrete dopant distribution in the channel," *J. of Computational Electronics*, vol. 5, no. 2–3, p. 119–123, Sept. 2006.
111. S. Sharma and P. Kumar, "Non overlapped single and double gate SOI/GOI MOSFET for enhanced short channel immunity," *J. of Semiconductor Technology and Science*, vol. 9, no. 3, pp. 136–147, Sept. 2009.
112. E. Bernard and O. Faynot, "First internal spacers introduction in record high I_{on}/I_{off} TiN/HfO$_2$ gate multichannel MOSFET satisfying both high performance and low standby power requirements," *IEEE Electron Device Letters*, vol. 30, no. 2, pp. 148–151, Feb. 2009.
113. T. C. Lim and G. A. Armstrong, "Scaling issues for analogue circuits using double gate SOI transistors," *Solid State Electronics*, vol. 51, no. 2, pp. 320–327, Feb. 2007.
114. Sungmo Kang and Yusuf Leblebichi, *CMOS Digital Integrated Circuits Analysis and Design*, 3rd Edition, McGraw-Hill, New York, USA, 2002.
115. Viranjay M. Srivastava, "Capacitance-voltage measurement for characterization of a metal-gate MOS process," *Int. J. of Recent Trends in Electrical and Electronics*, vol. 1, no. 4, pp. 4–7, May 2009.
116. D. Rechem, S. Latreche, and C. Gontrand, "Channel length scaling and the impact of metal-gate work function on the performance of double-gate MOSFETs," *J. of Physics*, vol. 72, no. 3, pp. 587–599, March 2009.
117. Usha Gogineni, Hongmei Li, Jesus Alamo, Susan Sweeney, and Basanth Jagannathan, "Effect of substrate contact shape and placement on RF characteristics of 45 nm low power CMOS devices," *IEEE J. of Solid State Circuits*, vol. 45, no. 5, pp. 998–1006, May 2010.
118. Usha Gogineni, Hongmei Li, Jesus Alamo, and Susan Sweeney, "Effect of substrate contact shape and placement on RF characteristics of 45 nm low power CMOS devices," *Proc. of Radio Frequency Integrated Circuits Symp.*, Boston, MA, USA, 7–9 June 2009, pp. 163–166.
119. Chien Ta, Efstratios Skafidas, and Robin Evans, "A 60 GHz CMOS transmit/receive switch," *Proc. of IEEE Radio Frequency Integrated Circuits Symp.*, Honolulu, Hawaii, USA, 3–5 June 2007, pp. 725–728.
120. Riza Tamer and Kausik Roy, "Analysis of options in double-gate MOS technology: A circuit perspective," *IEEE Trans. on Electron Devices*, vol. 54, no. 12, pp. 3361–3368, Dec. 2007.
121. Yiming Li and Hong Mu Chou, "A comparative study of electrical characteristic on sub 10-nm double-gate MOSFETs," *IEEE Trans. on Nanotechnology*, vol. 4, no. 5, pp. 645–647, May 2005.
122. Usha Gogineni, Jesus Alamo, and Christopher Putnam, "RF power potential of 45-nm CMOS technology," *Proc. of 10th Topical Meeting on Silicon Monolithic Integrated Circuits in RF Systems*, Phoenix, USA, 11–13 Jan. 2010, pp. 204–207.
123. Pedram Razavi and Ali Orouji, "Nanoscale triple material double gate MOSFET for improving short channel effects," *Proc. of Int. Conf. on Advances in Electronics and Microelectronics*, Valencia, USA, 29 Sept.–4 Oct. 2008, pp. 11–14.
124. Hamid Mahmoodi, Saibal Mukhopadhyay, and Kaushik Roy, "Design of high performance sense amplifier using independent gate control in sub 50 nm double-gate MOSFET," *Proc. of Int. Symp. on Quality Electronic Design*, San Jose, California, USA, 21–23 March 2005, pp. 490–495.
125. M. Steyaert, J. Janssens, B. Muer, M. Borremans, and N. Itoh, "A 2 V CMOS cellular transceiver front-end," *IEEE J. of Solid State Circuits*, vol. 35, no. 12, pp. 1895–1907, Dec. 2000.

References

126. Keunwoo Kim, Ching Te Chuang, Hung C. Ngo, Fadi H. Gebara, and Kevin J. Nowka, "Circuit techniques utilizing independent gate control in double-gate technologies," *IEEE Trans. on Very Large Scale Integration Systems*, vol. 16, no. 12, pp. 1657–1665, Dec. 2008.
127. C. Claeys, *ULSI Process Integration*, 1st Edition, Electrochemical Society Publications, New Jersey, USA, Jan. 2003.
128. Behzad Razavi, "A 300 GHz fundamental oscillator in 65-nm CMOS technology," *IEEE J. of Solid State Circuits*, vol. 46, no. 4, pp. 894–903, April 2011.
129. Changhwan Shin, Xin Sun, and King Liu, "Study of random dopant fluctuation effects for the tri-gate bulk MOSFET," *IEEE Trans. on Electron Devices*, vol. 56, no. 7, pp. 1538–1542, July 2009.
130. Giorgio Baccarani and Susanna Reggiani, "A compact double-gate MOSFET model comprising quantum-mechanical and nonstatic effects," *IEEE Trans. on Electron Devices*, vol. 46, no. 8, pp. 1656–1666, Aug. 1999.
131. Hugues Nurray, Patrik Martin, and Serge Bardy, "Taylor expansion of surface potential in MOSFET: application to Pao-Sah integral," *Active and Passive Electronic Components*, vol. 2010, pp. 1–11, 2010.
132. Jente B. Kuang, Keunwoo Kim, Ching Te Chuang, Hung C. Ngo, Fadi H. Gebara, and Kevin J. Nowka, "Circuit techniques utilizing independent gate control in double-gate technologies," *IEEE Trans. on Very Large Scale Integration Systems*, vol. 16, no. 12, pp. 1657–1665, Dec. 2008.
133. Kaushik Roy, Hamid Mahmoodi, and Saibal Mukhopadhyay, "Double-gate SOI devices for low-power and high-performance applications," *Proc. of 19th Int. Conference on VLSI Design*, India, 3–7 Jan., 2006, pp. 445–452.
134. M. Weis, R. Emling and D. Schmitt, "Circuit design with independent double-gate transistors," *J. Advances in Radio Science*, vol. 7, pp. 231–236, 2009.
135. Viranjay M. Srivastava, K. S. Yadav, and G. Singh, "Explicit model of cylindrical surrounding double-gate MOSFETs," *WSEAS Trans. on Circuits and Systems*, vol. 12, no. 3, pp. 81–90, March 2013.
136. Viranjay M. Srivastava, K. S. Yadav, and G. Singh, "DP4T RF CMOS switch: A better option to replace SPDT switch and DPDT switch," *Recent Patents on Electrical and Electronic Engineering*, vol. 5, no. 3, pp. 244–248, Oct. 2012.
137. Neil, Weste and David Harris, "*CMOS VLSI Design: A Circuits and Systems Perspective*," Pearson Addison Wesley, 3rd Edition, 2005.

Chapter 4
Double-Pole Four-Throw RF Switch Based on Double-Gate MOSFET

4.1 Introduction

In this chapter, we have designed a double-pole four-throw radio-frequency switch using double-gate (DP4T DG RF) MOSFET to operate at 0.1 GHz to few GHz frequency range for the advanced wireless communication systems. This switch mitigates attenuation of passing signals and exhibits high isolation to avoid misleading of simultaneously received signals. The symmetric DG MOSFET has been the focus of much attention for the application of RF switch due to its ability of strength to short-channel effects and improved current driving capability as discussed in the previous chapters.

4.2 Basics of Radio System Design

The RF system design is to connect various modules (e.g., amplifiers, oscillators, and mixers) in order to form a transceiver system [1]. To design this RF system, we required the components that work at RF range such as the baseband functionality, power supply, and antenna. In the designing of the DP4T RF switch, we have to use the following parameters:

4.2.1 Path Loss

To transmit a signal from A_Tx to A_Rx, it is necessary to transmit a signal of adequate power. The power of a signal of wavelength λ at a distance y in free space is given by Path Loss $= 20\log\left(\frac{4\pi y}{\lambda}\right) dB$. In addition to the free-space path loss, the signal is also attenuated by obstructions such as vegetation, buildings, and hills [2]. It should be noted that free-space path loss increases with increase in the frequency. So the higher frequency bands are generally used for shorter range or the line-of-sight communications.

In order to calculate a link budget, the power of the transmitter and the receiver's sensitivity are added to their antenna gains and to the path loss. For example, consider a transmitter of power 10 dBm transmitting through a directional antenna of gain 7 dBi to a receiver with 0 dBi antenna. The maximum path loss that this link can suffer is the sum of the antenna gains, and the difference between the transmit power and receive sensitivity, in this case, is 127 dB. It can be calculated that in free space, this would be equivalent to a range of 35 km at a frequency of 1 GHz. However, for a typical mobile radio environment, the range would be greatly reduced.

In addition to path loss, the factors such as reflection, diffraction, and interference affect the signal quality. The antenna is a particularly important factor as its height can be increased to see over obstacles. A high-gain antenna can be employed to combat path loss in a certain direction or to avoid the receiving interference from another direction. Two or more antennas can be used in a diversity scheme where the receiver (or transmitter) chooses to use the antenna with the best quality signal at a particular moment in time. So to connect these antennas, we have used DP4T switches in this book.

4.2.2 Gain Cascade

In the transceiver system, to work with gain described in decibels and power in dBm (decibel relative to 1 mw) is easier as these can simply be added and subtracted together when connecting system is in a chain.

4.2.3 1 dB Compression Point

A useful measure of the amount of power that a device can produce is 1 dB compression point [3, 4]. At low signal levels, a device is considered linear, but as the input signal is increased, the amount of signal at the output will begin to tail off. When the difference between the input and output reaches dB, the input power or output power is measured and referred as the input 1 dB point or output 1 dB point (Fig. 4.1).

4.2 Basics of Radio System Design

Fig. 4.1 1 dB compression point

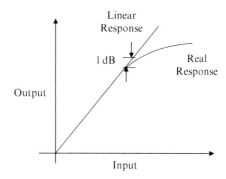

Fig. 4.2 Third-order intercept point

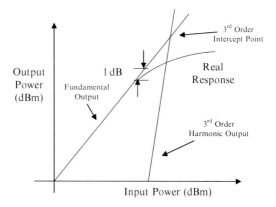

4.2.4 Third-Order Intercept Point

The third-order intercept point is a measure of linearity, which describes the amount of third-order harmonic in a device. These harmonics are produced in device due to the clipping of signal amplitude [5]. Third-order products are important as they fall near to the required frequency. In order to measure the third-order harmonics, two tones f_1 and f_2 are applied to the input, and the third-order harmonics can be viewed on a spectrum analyzer as shown in Fig. 4.2.

The level of fundamental and third-order output is plotted, and the two lines are extended to the theoretical point where they would cross is known as third-order intercept point. It should be noted that the third-order intercept point (IP_3) is theoretical and will change slightly depending on the signal level from which it is extrapolated.

4.2.5 Thermal Noise

The noise power from a resistor or resistive source is defined as the thermal noise floor in a given bandwidth. A useful number to remember is that at a temperature of

290 K (room temperature), the thermal noise floor is −204 or −174 dBm in a 1 Hz bandwidth $N = 10 \log(KTB)$ dBW, where $K = 1.38 \times 10^{-23}$ Boltzmann's constant, T is temperature in kelvin, and B is bandwidth in hertz [6, 7].

4.2.6 Noise Figure

The noise figure is a measure of the amount by which a device increases the noise power. An alternative method of expressing the noise introduced by a device is to give its equivalent noise temperature [8]. It can be used for modules such as filters and passive mixers which do not have gain if their loss in decibels is used in place of their noise figure.

4.2.7 Phase Noise

The phase noise from a local oscillator mixes with the required modulated signal during frequency mixing in the transmitter and receiver. One of the results is that in the frequency domain, in addition to its modulation sidebands, the required signal also has phase noise sidebands. The effect can be seen in the time domain as a phase error on the modulation. However, the poor phase noise performance can also give rise to undesirable levels of unwanted adjacent-channel power being radiated by transmitters and poor adjacent-channel selectivity in the receivers.

4.3 Design of DP4T DG RF CMOS Switch

We have designed a DP4T RF CMOS switch using DG MOSFET as shown in Fig. 4.3 for low power consumption and low distortion for RF switch in the communication that operates between 1 and 60 GHz [9]. This switch exhibit high isolation to avoid mixing of received signals at a particular time. This switch contains n-type DG MOSFET in its main architecture as shown in Fig. 4.3. In this switch, transmitted signal from power amplifier is sent to the transmitter through p-type DG MOSFET to transmitter named as A_T_x and B_T_x. Here, the two identical p-type DG MOSFETs are used and received signal which travel from the antenna through n-type DG MOSFET to the receiver named as A_R_x and B_R_x, two identical n-type DG MOSFETs. The switch designed in this chapter is suitable to drive 50 Ω resistive loads and utilizes multiple gate fingers to reduce the parasitic capacitance. The signal fading effects can be reduced with this DP4T DG RF CMOS switch because sending identical signals through the multiple antennas, which most likely provides the high-quality combined signal at the receiver end. The DG MOSFET as shown in Fig. 3.1 has two gates: G_1 (front gate) and G_2 (back gate).

4.3 Design of DP4T DG RF CMOS Switch

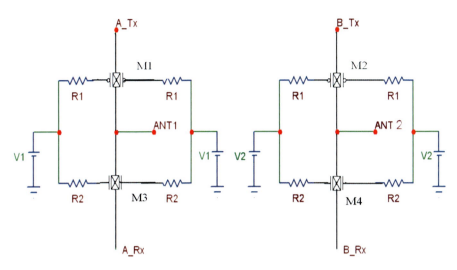

Fig. 4.3 Proposed DP4T DG RF CMOS switch

The back channel has been probed by varying the substrate (back gate) bias with the front-gate voltage as a parameter. These drain currents with respect to the gate-2 voltage, the curves are same as front-channel characteristics. The effective mobility in the front and back channels is comparable. In the regular silicon-on-insulator (SOI) MOSFETs, the buried oxide is much thicker (0.4 µm); hence, the substrate depletion has a minor effect even on the back channel. Moreover, the front channel is hardly affected because it is protected by the very small ratio between the thicknesses of the front and back oxides. In the present devices however, the substrate effect is exacerbated for several reasons [10]:

a. The oxide thickness is relatively thin.
b. The front and back oxides have equivalent thickness.
c. The ultrathin Si-film maximizes the coupling effects.

The transconductance decreases more rapidly with gate voltage in DG MOSFET, which indicates a larger value of the mobility degradation factor. Also, the transconductance gain of SG and DG at 300 °K tends to decrease in shorter channels. The gain of SG and DG in transconductance peaks (i.e., field-effect mobility) increases at the lower temperature as acoustic phonon scattering is gradually attenuated.

In Fig. 2.7, four transistors are used for two antennas. In this transceiver system using the CMOS functionality, at a time, any one of transistor M_1 (n-MOSFET) or M_3 (p-MOSFET) will operate, and in the same fashion, any one of transistor M_2 (n-MOSFET) or M_4 (p-MOSFET) will operate. However, the same working functions have been observed in this DP4T DG RF CMOS switch as in Fig. 4.3 [11]. Since the drain current consumption is significantly reduced, so the CMOS-based RF switches allow longer battery life than PIN diodes and about

60 % smaller as compared to smallest GaAs RF switch. Furthermore, this switch also experiences minimal distortion, negligible voltage fluctuation, and low power supply of only 1.2 V as shown in the following sections.

This designed switch is a part of the microwave applications for the switching system between transmitting and receiving modes. In this work, the designed DP4T switch has a symmetrical structure of transmitter (T_x) and receiver (R_x) to be operated at GHz range. The transistors M_1 and M_2 perform the main ON and OFF switching function, while the shunt transistors M_3 and M_4 which are cascade to M_1 and M_2, respectively, are used to improve the isolation of the switch by grounding RF signals on the side which is turned OFF. The switch can also connect coupling capacitors C_1 and C_2 which allow DC biasing of the T_x and R_x nodes of the switch. The gate resistances R_1 and R_2 are implemented to improve DC isolation. After designing of DP4T DG RF switch with designed DG MOSFET at 45-nm technology, we have drawn the layout and simulated the electrical parameters of this switch. It includes the basics of the circuit elements parameter required for the radio-frequency subsystems of the integrated circuits such as drain current, threshold voltage, resonant frequency, return loss, transmission loss, VSWR, resistances, capacitances, and switching speed.

4.4 Characteristics of DP4T DG RF CMOS Switch

For the purpose of current and switching speed, we have drawn the layout of DP4T DG CMOS transceiver switch with two input voltages (Vin_1 and Vin_2) and output through antennas (ANT_1 and ANT_2) with two transmitters (T_x_A and T_x_B) and two receivers (R_x_A and R_x_B) as shown in Fig. 4.4. Here the color codes have their usual meaning [12]. The layout is drawn for high-speed RF switch with the MOSFET channel width of 600 nm and channel length of 120 nm. This sizing ratio of the width and length can be scaled according to the application. These designs have a poly, drain, and source. The resistances are also present in this layout due to the metal connection with output voltage. The drain and source has equal capacitance of 0.19 fF, resistance of 90 Ω, and thickness of 2 μm, with metal capacitance of 0.13 fF, diffusion capacitance 0.06 fF, and gate capacitance of 0.86 fF. Here, we have analyzed the performance of DP4T DG CMOS switch by applying a gate voltage 1.2 V as low level. The start time, rise time, fall time, and pulse time for this signal is taken as 0.475 ns, 0.025 ns, 0.025 ns, and 0.475 ns, respectively. In this simulation, V_1 is 1.2 V as high level and start time, rise time, fall time, and pulse time for this signal are taken as 0.600 ns, 0.025 ns, 0.025 ns, and 0.475 ns, respectively, as shown in Fig. 4.5a. Assuming that MOSFET has symmetric structure and voltage applied on both gate are same this DP4T transceiver switch. Figure 4.5b shows the antenna voltages ANT_1 and ANT_2 with input voltages Vin_1 and Vin_2 for this transceiver switch. The drain current for this transceiver switch with output voltage is shown in Fig. 4.5c, which provides the drain current

4.4 Characteristics of DP4T DG RF CMOS Switch

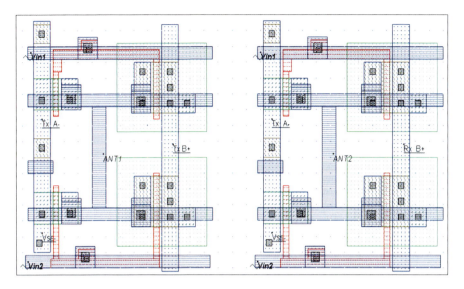

Fig. 4.4 Layout of the proposed DP4T DG RF CMOS switch

$I_{dd(max)} = 0.387$ mA, $I_{dd(avg)} = 0.02$ mA, and also rise time 36 ps at 1 GHz operating frequency.

In terms of antenna, we have drawn the first antenna (ANT$_1$) voltage at various frequencies from 0.1 to 8 GHz, in which the highest voltage on antenna is achieved at 1 GHz frequency and that decreases with the increase of frequency as shown in Fig. 4.5d. Also, this result is same for the second antenna (ANT$_2$). We have established this for the antenna at frequency of 1 GHz; the maximum data in terms of the voltage are observed. This data density decreases with the increase in the frequency. For example, at 0.93 GHz, voltage received by a transceiver system is 313.25 mV, and at 8 GHz this is 0.02 mV as shown in Fig. 4.5d. However, this result is same for the second antenna (ANT$_2$) as shown in Fig. 4.5d. For 1 GHz the drain current for single-pole double-throw (SPDT) transceiver switch is $I_{dd(max)} = 0.116$ mA, and the rise time 19 ps as well as for double-pole double-throw (DPDT) switch the drain current is $I_{dd(max)} = 0.193$ mA, and the rise time 31 ps for same operating frequency. After the simulation of DP4T transceiver switch design, we have obtained the results for drain current and switching speed for the SPDT, DPDT, and DP4T transceiver switches, which are summarized in Table 4.1.

The switching speed of the transceiver switch is characterized by the rise time and fall time. In this simulation, the control voltages are 50 % of duty pulses with a high level of 1 V and a low level of 0 V; the input signal is a sinusoid waveform at 1 GHz. The rise time (from 10 to 90 % of maximum output swing) is 19 ps–36 ps and fall time (90 % down to 10 % of maximum output swing) is 3 ps, respectively. This fast switching speed is possible owing to the small switching transistors. This sub-nanosecond switching speed is much faster than the speed of GaAs switch operating at GHz frequencies, which is usually on the order of tens of nanoseconds.

92   4 Double-Pole Four-Throw RF Switch Based on Double-Gate MOSFET

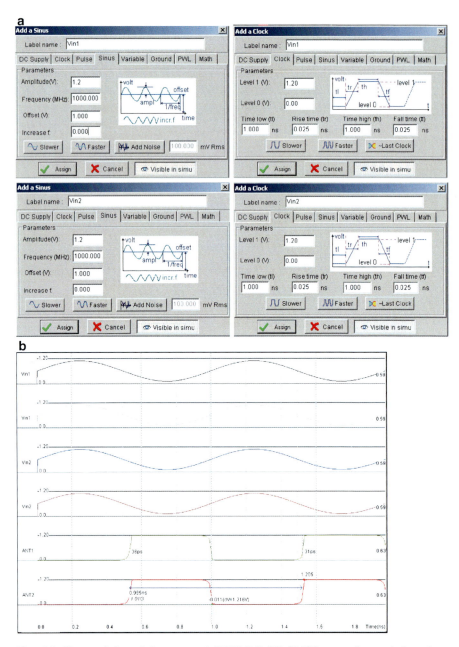

Fig. 4.5 Characteristics of the proposed DP4T DG RF CMOS transceiver switch such as (**a**) applied input voltages, (**b**) antenna voltage with input voltages, (**c**) drain current, and (**d**) antenna output at various frequencies

4.4 Characteristics of DP4T DG RF CMOS Switch

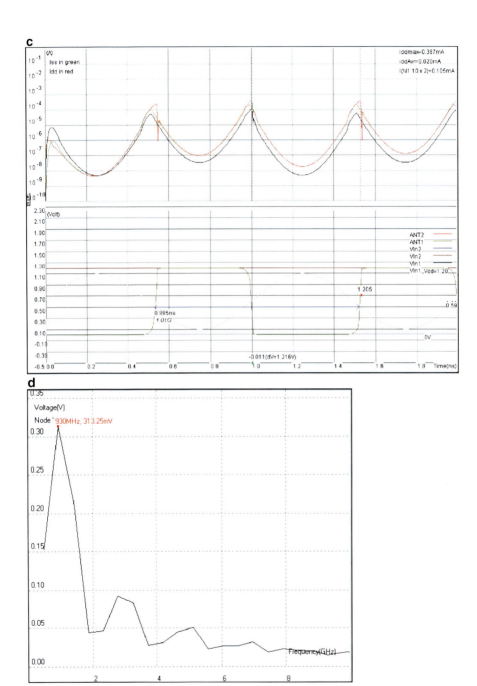

Fig. 4.5 (continued)

Table 4.1 Simulation results for drain current and switching speed for several switches

Structures	Drain current (mA)	Switching speed (ps)
SPDT	0.116	19
DPDT	0.193	30–31
DP4T	0.387	36

Table 4.2 Comparison of the switching speed

References	[9]	[13]	[14]	[15]
Switching speed	36 ps	300–1,000 ps	400 ps	40 ps

The switching speed can be improved by decreasing the values of the gate resistors as long as the resistors are still large enough to make the gates open to AC signal. The comparison of switching speed of the proposed DP4T transceiver switch is summarized in Table 4.2.

With the use of software or freely available tools for DG MOSFET length of 0.045 μm (45 nm) and width of 22.5 μm, we have calculated the capacitance of 5.72 fF, inductance of 30 pH, and resistance of 3.12 kΩ. After that, we designed a DP4T DG RF CMOS switch with the help of these values of the capacitance, inductance, and resistance. A small signal performance from few MHz to 100 GHz frequency range has been observed.

The equivalent switch circuit with respect to the capacitance is shown in Fig. 4.6 and the results are shown in Table 4.3 for the frequency band, noise power, delay, phase shift, return loss at T_x port, return loss at antenna port, insertion loss measured between T_x port and antenna port, voltage standing wave ratio (VSWR), transmission loss, reflection coefficient.

4.5 Effective ON-State Resistance of DP4T DG RF CMOS Switch

The models of an MOS transistor biased in deep triode region correspond to the ON-state of the switch, and cutoff region corresponds to the OFF-state of the switch. The insertion loss of a MOS transistor switch under the ON-state is dominated by its ON-state resistance (R_{ON}) and substrate/bulk resistance (R_b) [16–18]. However, the isolation of the switch under the OFF-state is finite due to the signal coupling through the parasitic capacitances, C_{ds}, C_{gs}, and C_{gd}, and through the junction capacitances, C_{sb} and C_{db}. Since the interface contact resistance is inversely proportional to the total gate area as in term of length and width of the gate. The reduction of resistance should lead to improved RF properties in MOSFETs [19–21]. The ON-state resistance is given by:

4.5 Effective ON-State Resistance of DP4T DG RF CMOS Switch

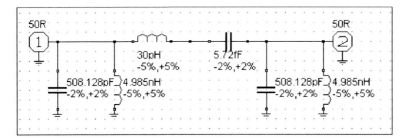

Fig. 4.6 Equivalent capacitive model of the proposed DP4T DG RF CMOS switch

Table 4.3 Performance parameters of the double-gate MOSFET transceiver switch

Parameter	Value and range
Frequency band	0.1–1 GHz
Temperature	27 °C
Noise power	4.4×10^{-21} W
Phase velocity	3×10^8 m/s
Delay	33.33 ps
Phase shift	1.2 degree at 0.1 GHz
Return loss	10 dB
Transmission loss	0.46 dB
VSWR	1.93:1
Reflection coefficient	0.316
RMS volts	0.707 V
Control voltages	1.2 V/0 V
I_{dd} max	0.387 mA
I_{dd} avg	0.020 mA
System impedance	50 Ω
Capacitance	31.83 pF
Inductance	79.88 nH
Reluctance	50 Ω^{-1}
t_{rise}	36 ps
t_{fall}	31 ps

$$R_{ON} = \frac{1}{\mu C_{ox} \frac{W}{L} (V_{gs} - V_{th})} \qquad (4.1)$$

For proper working of a switch and to reduce the insertion loss, we have to reduce this ON-state resistance. So to keep this R_{ON} small, we have to maintain the following steps.

4.5.1 Parallel Combination of Resistance in a Device

In DG MOSFET switch, there are two resistors between drain to source due to gate-1 and gate-2. However, both of these resistors are in parallel, so making the R_{ON} half, which keeps R_{ON} small as compared to the single-gate MOSFET. It means the switch can pass more signal as compared to the single-gate MOSFET-based switch.

4.5.2 Choosing Transistor with Large Mobility

For semiconductors, the behavior of transistors and other devices can be different depending on whether there are many electrons with low mobility or few electrons with high mobility. Therefore, the mobility is a very important parameter for semiconductor materials. However, the higher mobility leads to the better device performance, while the other parameters remain the same. This criterion of reducing R_{ON} is achieved by using n-type MOSFET transistors in place of p-type MOSFET transistors in the design [18].

4.5.3 Keeping $V_{gs} - V_{th}$ Large

This criterion is obtained by increasing the V_{gs} and decreasing the V_{th}, so that the difference of these is large. The threshold voltage has following equation:

$$V_{th} = V_{th_0} + \gamma \cdot \left(\sqrt{|2\phi_f + V_{sb}|} - \sqrt{|2\phi_f|} \right) \qquad (4.2)$$

where γ is the body effect coefficient and ϕ_f is the Fermi level of the substrate. Since in the double-gate MOSFET no bulk or substrate is available, γ equals to zero and this decreases the V_{th} which participates into the increasing of the $V_{gs} - V_{th}$. This leads to higher R_{ON} according to (4.1). Thus, by increasing the source/drain voltage, we sacrifice insertion loss for the power handling capability.

4.5.4 Aspect Ratio of a Transistor

To increase the ratio of W/L (aspect ratio), we have to widen the transistor width (W) and use the transistors of minimum allowable channel length (L). Since the minimum value of length is limited by the available technology, which is taken as 45 nm in this work. For this purpose, we have chosen the transistor of length 0.045 μm and width 22.5 μm. However, when we increase the width of a transistor, its junction capacitances and parasitic capacitances increase with the same ratio. For single-gate MOSFET, at ON condition of a transistor, increasing C_{sb} and C_{db} tends to have more signal being coupled with the substrate and dissipated in the substrate resistance R_b. At OFF-state of the transistor, increasing C_{ds}, C_{gd}, and C_{gs} tends to lower isolation between the sources and drain due to the capacitive coupling between these terminals [18, 22]. However, for DG MOSFET when both the transistors are ON, increasing C_{sb} and C_{db} leads to less signal being coupled to the substrate as substrate is not present in this structure, so there is no dissipation in the substrate resistance R_b. When the transistor is OFF, increasing C_{ds}, C_{gd}, and C_{gs} leads to higher isolation between the source and drain due to no capacitive coupling between these terminals.

For low frequency, sufficient isolation is achieved, so it is not an optimizing parameter for frequencies of order 1 GHz designs. Thus, in these designs, only insertion loss needs to be minimized, and there is no trade-off between the insertion loss and isolation required. On the other hand, at higher frequencies such as 60 GHz, isolation is smaller due to several low impedance paths caused by parasitic capacitances. This necessitates a trade-off between insertion loss and isolation during sizing the transistors. In this designed transistor, as the width increases for DG MOSFET, so peak power added efficiency and output power decrease. These are also because of a reduction in f_{max} [22].

4.6 Attenuation of DP4T CMOS Switch

The modern communication systems require variable attenuators and variable gain amplifiers for amplitude control in a variety of applications, such as automatic level control loops, modulators, and phased array systems. The variable attenuators are more suitable for applications, which require high linearity, low power consumption, and low temperature dependence, which cannot be achieved with variable amplifiers [23]. These attenuators mainly achieve relative attenuation from insertion loss differences by ON/OFF control of RF switches. The switched path attenuators use SPDT switches to steer the signal path between a line and a resistive network. This topology provides low phase variation over attenuation states, but it exhibits high insertion losses at reference states due to the cumulative losses of all SPDT switches for a multi-bit design, and it occupies a large chip area. Accordingly, it is not suitable for the design of CMOS digital step attenuators. The proposed design of DP4T RF CMOS switch overcomes this problem.

In this switch at ON-state condition, an effective resistance (R_{ON}) is estimated from input terminal to output terminal. Since R_{ON} changes with temperature (highest at high temperature), supply voltage, and to a minor degree with signal voltage and current. The ON-state resistance of CMOS switching elements can be approximated as follows:

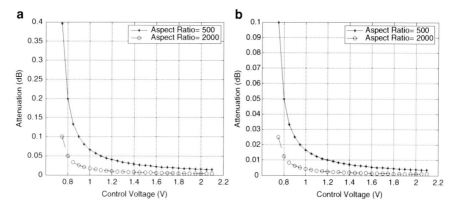

Fig. 4.7 Attenuation at $V_{CTL} = 0.7$–1.2 V for (**a**) 0.8-μm technology and (**b**) 45-nm technology

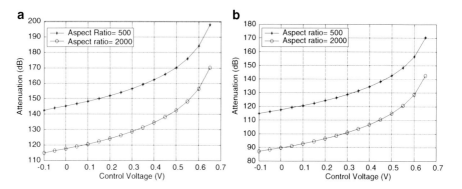

Fig. 4.8 Attenuation at $V_{CTL} = -0.1$ V to 0.7 V for (**a**) 0.8-μm technology and (**b**) 45-nm technology

$$R_{ON} = \frac{L}{W \cdot K_P \cdot (V_{CTL} - V_{th})} \quad (4.3)$$

where L, W, and V_{CTL} are the gate length (2.0, 1.2, or 0.8 microns), the gate width, and control voltage on the gate, respectively. The intrinsic transconductance K_P ranges from approximately 50 μA/V^2 for 2.0-μm technologies to 125 μA/V^2 for the 0.8-μm devices. The threshold voltages V_{th} for the devices are estimated to be 0.7 V–0.9 V range (for this work we have taken as 0.70 V). For a given value of ON-state resistance, a combination of larger transconductance K_P and smaller gate length L allows smaller gate widths W to be used, thereby significantly reducing the overall size of the microwave and RF switching transistors and hence the entire switching element.

The switching element was also tested in an attenuator configuration. The attenuation measurements have been performed on the 0.8-μm control element by varying the gate control voltage (V_{CTL}) over a 0 V–2.1 V range. Figures 4.7 and 4.8

4.6 Attenuation of DP4T CMOS Switch

shows the results of these measurements at 0.1 GHz, indicating a useful attenuation range up to 0.40 dB. The data are plotted with an attenuation model for comparison. The model is based on the use of the switching element as a series reflective attenuator. The level of attenuation (ATT) is given by the following expression [24]:

$$\text{ATT} = 20.\log\left(1 + \frac{R_{\text{ON}}}{2Z_0}\right) \quad (4.4)$$

where R_{ON} is the ON-state series resistance of the attenuator modeled using (4.3) and Z_0 is 50 Ω. The level of attenuation depends upon R_{ON}, with Z_0 (50 Ω). For lowering the attenuation, R_{ON} should be low and aspect ratio (W/L) should be high, which is achieved with the proposed DP4T DG RF CMOS switch as shown in Fig. 4.3 compared to DP4T SG RF CMOS switch as shown in Fig. 2.6b. So, we reached at the lower attenuation with the proposed switch at 45-nm technology.

The 50 Ω coaxial cables are the most commonly used coaxial cables, and they are used with radio transmitters, radio receivers, laboratory equipments, and in Ethernet network. Note that the minimal change in element characteristics over 0–5 V range is observed. For this purpose, at 0.8-μm technology DP4T DG RF CMOS switch design, we have compared the $L = 0.8$ μm and $W = 400$ μm with the $L = 2.0$ μm and $W = 4,000$ μm (where the aspect ratios are 500 and 2,000, respectively), and for 45-nm technology switch design, we compare $L = 0.045$ μm and $W = 22.5$ μm with $L = 0.045$ μm and $W = 90$ μm (where aspect ratios are 500 and 2,000, respectively, same as before). It is found that at lower size, its attenuation is lower [17, 22, 25]. The attenuation is estimated at the 0.8-μm and 45-nm technology by varying the gate control voltage (V_{CTL}) over a 0.0 V–2.1 V range. Figure 4.7a, b shows the results of evaluations at 1 GHz, indicating a useful attenuation range up to 0.4 dB and 0.1 dB. The data are plotted with an attenuation model for comparison.

From Fig. 4.7a, b at control voltage 1.0 V, attenuations for aspect ratio 500 are 0.070 dB and 0.016 dB at 0.8-μm and 45-nm technology, respectively. Similarly, the attenuations for aspect ratio 2,000 are 0.020 dB and 0.005 dB at 0.8-μm technology and 45-nm technology, respectively. Similarly, at the various control voltages from −0.1 V to 0.7 V for the DP4T DG RF CMOS switch as shown in Fig. 4.8, we observed that the attenuations for aspect ratio 500 are 199 dB and 170 dB at 0.8-μm and 45-nm technology, respectively. Similarly, the attenuations for aspect ratio 2,000 are 170 dB and 142 dB at 0.8-μm technology and 45-nm technology, respectively. These results of the attenuations are summarized in Tables 4.4 and 4.5.

From preceding discussions, we conclude that at higher technology, the attenuation is lower as compared to the lower technology attenuations. With scaling device dimensions and increasing short-channel effects, multiple gate transistors are investigated to obtain and improve the gate control. However, this design of double-gate transistors resolves the problem of short-channel effect occurrence in MOSFET structures [24–27]. If both the gates of double-gate are independently controlled, then logic density as well as logic functionality can be increased.

Table 4.4 DP4T DG RF CMOS switch attenuation for control voltage range 0.7 V–2.1 V

Technology	W (μm)	L (μm)	Aspect ratio	Attenuation (dB)
0.8 μm	400	0.8	500	0.070
	4,000	2.0	2,000	0.020
45 nm	22.5	0.045	500	0.016
	90	0.045	2,000	0.005

Table 4.5 DP4T DG RF CMOS switch attenuation for control voltage range –0.1 V to 0.7 V

Technology	W (μm)	L (μm)	Aspect ratio	Attenuation (dB)
0.8 μm	400	0.8	500	199
	4,000	2.0	2,000	170
45 nm	22.5	0.045	500	170
	90	0.045	2,000	142

4.6.1 Causes of Attenuation

However, both the signal frequency and the range between the end points of the medium affect the amount of attenuation. As either the frequency or the range increases, the attenuation increases. Unlike open outdoor applications based on straightforward free-space loss formulas, attenuation for indoor systems is very complex to calculate. The main reason for this difficulty is that the indoor signals bounce off obstacles and penetrate a variety of materials that offer varying effects on the attenuation [28].

4.6.2 Counteracting Attenuation

The main objective of warfare attenuation is to avoid having signal power within the area where users operate to fall below the sensitivity of the 802.11 radio receivers. We need to ensure that the receiver is always able to find the transmissions. However, the higher levels of RF interference, such as that caused by 2.4 GHz on Bluetooth devices or cordless phones, will negatively impact the ability of the receiver to decode the signal. As RF interference signal levels become higher than 802.11 signals, an 802.11 receiver will encounter considerable bit errors while trying to demodulate the 802.11 signals [29].

The mathematical method for determining acceptable attenuation is the equivalent isotropically radiated power (EIRP) and receiver sensitivity [30]. The receiver sensitivity is different and depending upon whether we use 802.11a or 802.11b and the data rate at which the users are operating [31, 32]. The higher data rate lowers the receiver sensitivity requirements. In other words, a receiver must be more sensitive to detect higher data rate signals. For example, the EIRP of the source station could be 200 mW (23 dBm) and the receiver sensitivity would be –76 dBm for 802.11b at 11 Mbps. Thus, one can only afford to have 99 dB of attenuation (23 dBm to −76 dBm) before the signal drops below the receiver's ability to hear the signal. The use of an 802.11 radio proves that the signal levels are above minimum requirements. Using a wireless local area network (WLAN) analyzer, as AirMagnet or AiroPeek, we can measure the signal power at various points to ensure that the signal power levels are well above the receiver sensitivity [33].

4.7 OFF-Isolation

If we apply a high-frequency signal to an open switch input appearing at the output, then OFF-isolation is as

$$\text{Off-isolation} = 20 \times \log\left(\frac{V_{in}}{V_{out}}\right) \qquad (4.5)$$

The signal is transmitted through capacitance between drain to source (C_{ds}) to a load composed of drain capacitance (C_d), which is parallel to the external load. This isolation decreases by 6 dB/octave with the rising frequency and depends on the threshold voltage. These T_{ON} and T_{OFF} includes the propagation delay of the logic signal through the logic driving circuits. The fast logic signal propagates through the internal circuitry of the switch before triggering the transmission line. It is also verified by the properties of DG CMOS with the high merit of design.

4.8 Resistive and Capacitive Model of DP4T DG RF CMOS Switch

The capacitive models of a DP4T DG RF CMOS switch is biased in linear region. For the ON-state of switch, at a time only one transistor will be ON either n-type DG MOSFET or p-type DG MOSFET. The capacitive model for n-type DG MOSFET is shown in Fig. 3.7a. For the given design of DP4T DG RF CMOS switch, the total maximum capacitance across source to drain model under the operating condition is shown in Fig. 4.9, which is calculated by (4.6). Here we consider the ON-state, because the insertion loss is conquered by its ON-state resistance and substrate/bulk resistance. The isolation of the switch is finite due to the signal coupling through the parasitic and junction capacitances. For DG MOSFET at cutoff region, the resistances R_{ON1} and R_{ON2} will be zero. For maximum capacitance, assuming all the capacitances is present at a time. In the DG MOSFET, the parasitic capacitances are C_{ds1}, C_{ds2}, C_{gs1}, C_{gs2}, C_{gd1}, and C_{gd2}, and junction capacitances are not present as bulk that is not available.

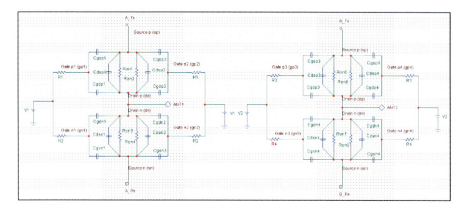

Fig. 4.9 Resistive and capacitive model of DP4T DG RF CMOS switch at ON-state

For DG MOSFET, when both the transistors are ON, C_{sb} and C_{db} are not present, so fewer signals being coupled to the substrate as substrate are not present in this structure, therefore no dissipation in the substrate resistance R_b. When the transistor is in cutoff region, the increase in C_{ds1}, C_{ds2}, C_{gd1}, C_{gd2}, C_{gs1}, and C_{gs2} leads to the higher isolation between the sources and drain due to without capacitive coupling between these terminals. At the transistors cutoff region, C_{ds}, C_{gd}, and C_{gs} increase, which directs to the lower isolation between the source and drain due to the capacitive coupling between these terminals [12]:

$$C_{DG} = C_{ds1} + C_{ds2} + \frac{C_{gs1} \cdot C_{gd1}}{C_{gs1} + C_{gd1}} + \frac{C_{gs2} \cdot C_{gd2}}{C_{gs2} + C_{gd2}} \quad (4.6)$$

For the model of a DP4T DG RF CMOS switch at ON-state, the total capacitance is individual capacitance of a particular transistor as any one transistor is ON at a time which is calculated for n-type DG MOSFET and p-type DG MOSFET as shown below, respectively:

$$C_{DGn} = C_{dsn1} + C_{dsn2} + \frac{C_{gsn1} \cdot C_{gdn1}}{C_{gsn1} + C_{gdn1}} + \frac{C_{gsn2} \cdot C_{gdn2}}{C_{gsn2} + C_{gdn2}} \quad (4.7)$$

$$C_{DGp} = C_{dsp1} + C_{dsp2} + \frac{C_{gsp1} \cdot C_{gdp1}}{C_{gsp1} + C_{gdp1}} + \frac{C_{gsp2} \cdot C_{gdp2}}{C_{gsp2} + C_{gdp2}} \quad (4.8)$$

and the ON-state resistance is parallel resistance due to the parallel combination of gate-1 (G_1) and gate-2 (G_2) as

$$R_{DG} = \frac{R_{ON1} \cdot R_{ON2}}{R_{ON1} + R_{ON2}} \quad (4.9)$$

where

$$R_{ON} = \frac{1}{\mu C_{ox} \frac{W}{L} (V_{gs} - V_{th})} \quad (4.10)$$

With the calculation of capacitances using (4.6) and (4.7), we conclude that if the value of capacitances is increased, then isolation is better in DP4T DG RF CMOS switch as compared to the already existing DP4T RF CMOS switch. Also, the resistance R_{ON} (parallel combination of R_{ON1} and R_{ON2}) is less which helps in fast current movement and increases the speed of switch. For appropriate working of a switch and to reduce the insertion loss, we can also achieve reduction in ON-state resistance with choosing transistor with large μ, increasing W/L, keeping $V_{gs}-V_{th}$ large as revealed from (4.10) as discussed in the Sect. 4.5.

4.9 Switching Speed of DP4T DG RF CMOS Switch

The switching speed is the measure of the rate at which a given electronic logic device is capable of changing the logic state of its output in response to the changes at its input. It is a function of the delay encountered within the device, which in turn is a function of the device technology. In an electronic switch, the measure of the switching transient is the 10–90 % rise time or fall time. This is the time taken for the signal to transition between 10 and 90 % of its total swing. If the waveform is exponential, then there is a relationship between the 10 and 90 % rise time and the time constant in the circuit. The time it takes to reach 10 % of the final value is $t_{10\%} = -RC \ln(0.9)$ and for 90 % of the final value is $t_{90\%} = -RC \ln(0.1)$. So, the 10–90 % rise time will be $T_{10-90\%} = t_{90\%} - t_{10\%} = -RC \ln(0.9/0.1) = 2.2\ RC$ and a similar result for the fall time. For the proposed DP4T DG RF CMOS switch, we put the values of $R = 12.5\ \Omega$ and $C = 1.4$ pF, which are obtained from (4.6) and (4.9); on this way, we have achieved the rise time of 36 ps which is justified with Fig. 4.5b.

4.10 S-Parameters of DP4T DG RF CMOS Switch

The S-parameters describe the response of an N-port network to voltage signals at each port. It has magnitude as well as phase information for gain or loss and phase shift information. However, the S-parameters change with the frequency, load impedance, source impedance, and network. For an example, we acquire each available capacitance in Fig. 4.9 of 1 fF for DP4T DG RF CMOS switch. It gives total capacitance of C_{DG} of a 3 fF with help of (4.6). With this capacitance, we have designed a switch network as shown in Fig. 4.10 [34, 35]. Here, the tolerance of total capacitance is considered as +2 % to −2 %. At frequency band of 0.8 GHz–60 GHz,

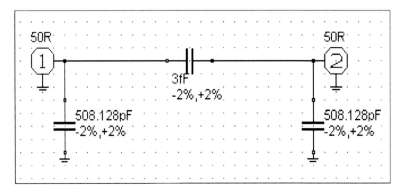

Fig. 4.10 Equivalent capacitive circuit of the DP4T DG RF CMOS switch

Table 4.6 Simulated parameters of the DP4T DG RF CMOS switch

Parameters	Measurements
Operating frequency	0.8–60 GHz
System impedance	50 Ω
Tolerance of capacitance	+2 % to −2 %
Noise power	−173.83 dBm
Return loss	10 dB
Transmission loss	0.46 dB
VSWR	1.93:1
Reflection coefficient	0.316

Table 4.7 Impedance, admittance, series equivalent, and parallel equivalent circuit parameters of the proposed switch

Freq. GHz	Impedance Real	Img. (j)	Admittance Real	Img. (j)	Series equivalent R (Ω)	C (pF)	Parallel equivalent R (Ω)	C (pF)
1	64.1f	−0.30	707.62f	+3.32	64.10f	508.13	1.41 T	508.13
2.5	10.32f	−0.13	647.69f	+7.92	10.32f	508.13	1.54 T	508.13
5	3.52f	−0.06	897.02f	+15.97	3.52f	508.13	1.11 T	508.13
10	1.09f	−0.03	1.13p	+32.08	1.09f	508.13	888.5G	508.13
25	$-8.27e^{-16}$	−0.01	−5.04p	+78.08	$-8.27e^{-16}$	508.13	−198.26G	508.13
50	$-1.76e^{-16}$	−6.26 m	−4.5p	+159.74	$-1.76e^{-16}$	508.13	−222.28G	508.13
60	$-1.37e^{-16}$	−5.22 m	−5.03p	+191.57	$-1.37e^{-16}$	508.13	−198.88G	508.13

we obtained the noise power of −173.83 dBm with return loss 10 dB, transmission loss 0.46 dB, VSWR 1.93:1, reflection coefficient 0.316, and system impedance 50 Ω. We have analyzed this DP4T DG RF CMOS equivalent switch for 1, 2.5, 5, 10, 25, 50, and 60 GHz of frequency [18, 36–40], and the parameters are summarized in Table 4.6. Table 4.7 shows the impedance, admittance, series equivalent, and parallel equivalent of this network at 1–60 GHz frequencies.

4.10 S-Parameters of DP4T DG RF CMOS Switch

Table 4.8 S-parameters of a designed switch at various frequencies (Mag. = magnitude, Ang. = angle)

Frequency GHz	S_{11} Mag.	S_{11} Ang.	S_{21} Mag.	S_{21} Ang.	S_{12} Mag.	S_{12} Ang.	S_{22} Mag.	S_{22} Ang.
1	1	−179.22	0	−89.22	0	−89.22	1	−179.22
2.5	1	−179.71	0	−89.71	0	−89.71	1	−179.71
5	1	−179.86	0	−89.86	0	−89.86	1	−179.86
10	1	−179.93	0	−89.93	0	−89.93	1	−179.93
25	1	−179.97	0	−89.97	0	−89.97	1	−179.97
50	1	−179.98	0	−89.98	0	−89.98	1	−179.98

Table 4.9 Magnitude of S_{12} and S_{21} (both are equal) at various frequencies

Frequency GHZ	Magnitude dB=20·log$_{10}$ (V)
1	−142.97
2.5	−150.51
5	−156.60
10	−162.66
25	−170.60
50	−176.60
60	−178.18

After designing this network of switch, we tabularize the S_{11}, S_{12}, S_{21}, and S_{22} in Table 4.8, at the above-stated frequencies. These values of S-parameters give the magnitude of S_{12} and S_{21} as shown in Table 4.9.

From Table 4.7, it is revealed that the impedance, admittance, series equivalent resistance, and parallel equivalent resistance decrease with the increase in frequency, but the series equivalent capacitance and parallel equivalent capacitance are fixed at 508.13 pF, whatever the value of frequency. From Table 4.8, it has been concluded that at no signal, the entire unpredictable signal returned back means no any unwanted signal passes the switch because S_{11} and S_{22} are 1 and at the same time S_{12} and S_{21} are zero. Also with increase in frequency, we achieve a fine angle.

From Table 4.8, it is illustrated that the maximum amplitude and minimum amplitude of the resultant signal occur when the reflected wave is in phase and 180° out of phase with the input signal, respectively. The reflected wave, when summed with the input signal, either increases or decreases its net amplitude, depending on whether the reflection is in phase or out of phase with the input signal.

From Table 4.9, it is accomplished that the magnitude in terms of dB decreases with increase in the frequency. We can also measure the reflection, VSWR (ratio of the maximum reflected wave in phase to minimum reflected wave out of phase voltages in the standing wave), return loss (measure of the reflected signal power), and also a subset of the insertion loss as the higher the return loss (or reflections) in an RF system, the higher its insertion loss from these conclusions. Due to the reduced short-channel effects, the double-gate SOI devices have emerged as the device of choice for nanoscaled technologies. The device characteristic including ON and OFF currents can be optimized by the choice of device geometries, aspect ratios, gate material, work function, threshold voltages, etc.

However, the possibilities of independent double-gate technologies (four-terminal devices) have significant advantage for low-power and high-performance circuit design [14, 15, 41]. The usefulness of the proposed design of SPDT, DPDT, and DP4T switch is shown in a class of low-power and high-performance circuits such as wireless communication system, dynamic logic circuits, Schmitt trigger, SRAM cells, and sense amplifiers [42].

4.11 Conclusions

From the preceding analysis and simulations, we have obtained that the attenuation of a DP4T DG CMOS switch decreases as compared to the single CMOS switch. This switch is highly integratable with the analog and digital circuitry. These features make DP4T DG CMOS switch very attractive for use in applications that require mixed RF and digital systems. The proposed DP4T DG RF CMOS switch exhibits better drain current and switching speed compared to existing SPDT and DPDT switch. The insertion loss and power handling capability of the switch is also improved by using a higher control voltage of 1.2 V. However, based on the simulation results, it is demonstrated that the drain currents increase in a way from SPDT to DPDT and DPDT to DP4T switch structures. A better reduction of the short-channel effects and improvement of the device reliability could also be expected by changing the channel width and channel length ratio and gate materials and optimizing these parameters. The proposed DP4T DG RF CMOS switch results the peak output currents around 0.387 mA and switching speed of 36 ps. A device structure with a double-gate contact shows a significant improvement in currents and switching speed compared to a single-gate contact structure. To reduce the short-channel effects of nanoscale devices, the DG MOSFET can also be designed using this process. This switch is suitable for low-power short-range wireless communication devices.

In the DG MOSFET, the bulk voltage is zero, so we can achieve the highest drain current by using the proposed switch. At higher technology, attenuation is lower as we reported 0.01 dB to 0.05 dB for 45-nm technology compared to 0.05 dB–0.20 dB for 0.8-µm technology in our work [27, 43]. OFF-isolation and switching speed are significantly improved in the proposed DP4T DG RF CMOS switch over the already existing CMOS switch. Moreover, the flat-band capacitance and power dissipation become half and threshold voltage as well as flat-band voltages is reduced as flat-band capacitance becomes half for the proposed DP4T DG RF CMOS switch. Ultrathin body employs very thin silicon body to achieve better control of the channel by the gate and, hence, reduces the leakage and short-channel effects. The intrinsic or lightly doped body, in the DG MOSFET, reduces the threshold voltage variations due to random dopant fluctuations and enhances the mobility of the careers in the channel region which increases the ON-state current. So, we can achieve a better result by using this DG MOSFET at 45-nm technology as it has intrinsic or lightly doped body for the application of DP4T DG RF CMOS switch.

References

1. M. Hunter, The basics of radio system design: how to design RF circuits? *IEE Training Course*, vol. 10, no. 1, p. 17, 2000.
2. K. T. Herring, J. W. Holloway, D. H. Staelin, and D. W. Bliss, "Path-loss characteristics of urban wireless channels," *IEEE Trans. on Antennas and Propagation*, vol. 58, no. 1, pp. 171-177, Jan. 2010.
3. S. Chartier, P. Lohmiller, J. Dederer, H. Schumacher, and M. Oppermann, "SiGe BiCMOS wideband low noise amplifiers for application in digital beam-forming receivers," *Proc. European Microwave Conference*, Paris, 28–30 Sept. 2010, pp.1070–1073.
4. Q. Jie, J. D. Cali, B. F. Dutton, G. J. Starr, D. Fa Foster, and C. E. Stroud, "Selective spectrum analysis for analog measurements," *IEEE Trans. on Industrial Electronics*, vol. 58, no. 10, pp. 4960-4971, Oct. 201.1
5. Heng Zhang and Sanchez Sinencio, "Linearization techniques for CMOS low noise amplifiers: A tutorial," *IEEE Trans. on Circuits and Systems*, vol. 58, no. 1, pp. 22-36, Jan. 2011.
6. M. Emam, P. Sakalas, D. Janvier, J. P. Raskin, L. T. Chuan, and F. Danneville, "Thermal noise in MOSFETs: A two- or a three-parameter noise model?," *IEEE Trans. on Electron Devices*, vol. 57, no. 5, pp. 1188-1191, May 2010.
7. S. Ickhyun, Jeon Jongwook, Hee Sauk, Kim, Junsoo Park, Byung Gook, Jong Duk Lee, and Hyungcheol Shin, "A simple figure of merit of RF MOSFET for low-noise amplifier design," *IEEE Electron Device Letters*, vol. 29, no. 12, pp. 1380–1382, Dec. 2008.
8. A. J. Lelis, D. Habersat, R. Green, A. Ogunniyi, M. Gurfinkel, J. Suehle, and N. Goldsman, "Time dependence of bias-stress-induced SiC MOSFET threshold-voltage instability measurements, " *IEEE Trans. on Electron Devices*, vol. 55, no. 8, pp. 1835–1840, Aug. 2008.
9. Viranjay M. Srivastava, K. S. Yadav, and G. Singh, "Design and performance analysis of double-gate MOSFET over single-gate MOSFET for RF Switch," *Microelectronics Journal*, vol. 42, no. 3, pp. 527–534, March 2011.
10. A. O. Adan, M. Koyanagi, and M. Fukumi, "Physical model of noise mechanisms in SOI and bulk-silicon MOSFETs for RF applications," *IEEE Trans. on Electron Devices*, vol. 55, no. 3, pp. 872–880, March 2008
11. Etienne Sicard and Sonia Delmas Bendhia, *Basics of CMOS Cell Design*, 1st Edition, McGraw-Hill, USA, 2005.
12. Sungmo Kang and Yusuf Leblebichi, *CMOS Digital Integrated Circuits Analysis and Design*, 3rd Edition, McGraw-Hill, New York, USA, 2002.
13. A. M. Street, *RF Switch Design, IEE Training Course*, United Kingdom, vol. 4, pp. 1–7, April 2000.
14. Kwangchun Jung and K. O. Kenneth, "A CMOS single-pole four-throw switch," *IEEE Microwave and Wireless Components Letters*, vol. 16, no. 3, pp. 128–130, March 2006.
15. Yuan Taur, Wei Chen, D. J. Frank, Shih Hsien Lo, and Hon Sum Wong, "CMOS scaling into the nanometer regime," *Proc. of IEEE*, vol. 85, no. 4, pp. 486–504, Aug. 2002.
16. F. J. Huang and O. Kenneth, "A 0.5 μm CMOS T/R switch for 900 MHz wireless applications," *IEEE J. of Solid State Circuits*, vol. 36, no. 3, pp. 486–492, March 2001.

17. Usha Gogineni, Hongmei Li, Jesus Alamo, and Susan Sweeney, "Effect of substrate contact shape and placement on RF characteristics of 45 nm low power CMOS devices," *Proc. of Radio Frequency Integrated Circuits Symp.*, Boston, MA, USA, 7–9 June 2009, pp. 163–166.
18. Chien Ta, Efstratios Skafidas, and Robin Evans, "A 60 GHz CMOS transmit/receive switch," *Proc. of IEEE Radio Frequency Integrated Circuits Symp.*, Honolulu, Hawaii, USA, 3–5 June 2007, pp. 725–728.
19. Behzad Razavi, "A 300 GHz fundamental oscillator in 65-nm CMOS technology," *IEEE J. of Solid State Circuits*, vol. 46, no. 4, pp. 894–903, April 2011.
20. S. Chouksey and J. G. Fossum, "DICE: a beneficial short-channel effect in nanoscale double-gate MOSFETs," *IEEE Trans. on Electron Devices*, vol. 55, no. 3, pp. 796–802, March 2008.
21. Viranjay M. Srivastava, K. S. Yadav, and G. Singh, "DP4T RF CMOS switch: A better option to replace SPDT switch and DPDT switch," *Recent Patents on Electrical and Electronic Engineering*, vol. 5, no. 3, pp. 244–248, Oct. 2012.
22. Viranjay M. Srivastava, K. S. Yadav, and G. Singh, "Analysis of Drain Current and Switching Speed for SPDT Switch and DPDT Switch with Proposed DP4T RF CMOS Switch," *J. of Circuits, Systems, and Computers*, vol. 21, no. 4, pp. 1–18, June 2012.
23. Byung W. Min and Gabriel M. Rebeiz, "A 10 to 50 GHz CMOS distributed step attenuator with low loss and low phase imbalance," *IEEE J. of Solid State Circuits*, vol. 42, no. 11, pp. 2547–2554, Nov. 2007.
24. Viranjay M. Srivastava, K. S. Yadav, and G. Singh, "Attenuation with double pole four throw CMOS switch design," *Proc. of IEEE Int. Conf. on Semiconductor Electronics*, Malaysia, 28–30 June 2010, pp. 173–175.
25. K. Raczkowski, S. Thijs, W. De Raedt, B. Nauwelaers, and P. Wambacq, "50 to 67 GHz ESD protected power amplifiers in digital 45 nm LP CMOS," *Proc. of Int. Conf. on Solid State Circuits*, San Francisco, California, USA, 8–12 Feb. 2009, pp. 382–384.
26. Usha Gogineni, Hongmei Li, Jesus Alamo, Susan Sweeney, and Basanth Jagannathan, "Effect of substrate contact shape and placement on RF characteristics of 45 nm low power CMOS devices," *IEEE J. of Solid State Circuits*, vol. 45, no. 5, pp. 998–1006, May 2010.
27. A. Valdes Garcia, S. Reynolds, and J. O. Plouchart, "60 GHz transmitter circuits in 65 nm CMOS," *Proc. of Radio Frequency Integrated Circuits Symp.*, Atlanta, 15–17 June 2008, pp. 641–644.
28. Viranjay M. Srivastava, K. S. Yadav, and G. Singh, " Performance of double-pole four-throw double-gate RF CMOS Switch in 45 nm technology," *Int. J. of Wireless Engineering and Technology*, vol. 1, no. 2, pp. 47–54, Oct. 2010.
29. S. H. Lee, C. S. Kim, and H. K. Yu, "A small signal RF model and its parameter extraction for substrate effects in RF MOSFETs," *IEEE Trans. on Electron Devices*, vol. 48, no. 7, pp. 1374–1379, July 2001.
30. G. Chien, F. Weishi, Y. Hsu, and L. Tse, "A 2.4 GHz CMOS transceiver and baseband processor chipset for 802.11b wireless LAN application," *Proc. of IEEE Int. Conf. on Solid State Circuits*, San Francisco, 9–13 Feb. 2003, pp. 358–359.
31. W. Kluge, L. Dathe, R. Jaehne, and D. Eggert, "A 2.4 GHz CMOS transceiver for 802.11b wireless LANs," *Proc. of IEEE Solid State Circuits Conf.*, San Francisco, California, USA, 9–13 Feb. 2003, pp. 360–361.
32. D. Su, M. Zargari, P. Yue, D. Weber, B. Kaczynski, and B. Wooley, "A 5 GHz CMOS transceiver for IEEE 802.11a wireless LAN systems," *IEEE J. of Solid State Circuits*, vol. 37, no. 12, pp. 1688–1694, Dec. 2002.
33. D. Su, M. Zargari, P. Yue, D. Weber, B. Kaczynski, and B. Wooley, "A 5 GHz CMOS transceiver for IEEE 802.11a wireless LAN," *Proc. of IEEE Int. Conf. on Solid State Circuits*, San Francisco, California, USA, 7 Feb. 2002, pp. 92–93.
34. S. Sharma and P. Kumar, "Non overlapped single and double gate SOI/GOI MOSFET for enhanced short channel immunity," *J. of Semiconductor Technology and Science*, vol. 9, no. 3, pp. 136–147, Sept. 2009.

References

35. Thomas H. Lee, *Planar Microwave Engineering: A Practical Guide to Theory, Measurement and Circuits*, 2nd Edition, Cambridge University Press, India, 2006.
36. C. Y. Lin, "Design and implementation of configurable ESD protection cell for 60 GHz RF circuits in a 65 nm CMOS process," *Microelectronics Reliability*, vol. 51, no. 8, pp. 1315–1324, Aug. 2011.
37. J. P. Carmo, P. M. Mendes, C. Couto, and J. H. Correia, "A 2.4 GHz RF CMOS transceiver for wireless sensor applications," *Proc. of Int. Conf. on Electrical Engineering*, Coimbra, Portugal, 2005, pp. 902–905.
38. P. H. Woerlee, M. J. Knitel, and A. J. Scholten, "RF CMOS performance trends," *IEEE Trans. on Electron Devices*, vol. 48, no. 8, pp. 1776–1782, Aug. 2001.
39. Viranjay M. Srivastava, K. S. Yadav, and G. Singh, "Analysis of double gate CMOS for double-pole four-throw RF switch design at 45-nm technology," *J. of Computational Electronics*, vol. 10, no. 1–2, pp. 229–240, June 2011.
40. Y. Cheng and M. Matloubian, "Parameter extraction of accurate and scaleable substrate resistance components in RF MOSFETs," *IEEE Electron Device Letters*, vol. 23, no. 4, pp. 221–223, April 2002.
41. Li Zhiyuan, Ma Jianguo, Ye Yizheng, and Yu Mingyan, "Compact channel noise models for deep-submicron MOSFETs," *IEEE Trans. on Electron Devices*, vol. 56, no. 6, pp. 1300–1308, June 2009.
42. A. Poh and Z. Ping, "Design and analysis of transmit/receive switch in triple-well CMOS for MIMO wireless systems," *IEEE Trans. on Microwave Theory and Techniques*, vol. 55, no. 3, pp. 458–466, March 2007.
43. S. Solda, M. Caruso, A. Bevilacqua, A. Gerosa, D. Vogri, and A. Neviani, "A 5 Mbps UWB-IR transceiver front-end for wireless sensor networks in 0.13 μm CMOS," *IEEE J. of Solid State Circuits*, vol. 46, no. 7, pp. 1636–1647, July 2011.

Chapter 5
Cylindrical Surrounding Double-Gate RF MOSFET

5.1 Introduction

To reduce the size and increase the compactness in terms of area for the designed double-gate (DG) MOSFET as discussed in Chap. 3, we have analyzed and model the gate all around to the DG MOSFET. In this chapter we have designed the cylindrical surrounding double-gate (CSDG) MOSFET and analyzed the design parameters of this MOSFET as a RF switch for the advanced wireless telecommunication systems. We have emphasized on the basics of the circuit parameters such as drain current, threshold voltage, resonant frequency, resistances at switch ON-state condition, capacitances, energy stored, cross talk, and switching speed required for the integrated circuit of the radio frequency subsystem of the CSDG MOSFET device and physical significance of these basic circuit parameters are also discussed. We have analyzed that CSDG MOSFET stored more energy (1.4 times) as compared to the cylindrical surrounding single-gate (CSSG) MOSFET. The ON-state resistance of CSDG MOSFET is half as compared to the DG MOSFET and SG MOSFET, which reveals that the current flow from source to drain in CSDG MOSFET is better than that of the DG MOSFET and SG MOSFET.

There has been growing interest in the modeling of RF CMOS because it allows integration with both the digital and analog functionality on the same die, means increasing the component density with increasing performance at the same time as keeping system sizes reserved [1, 2]. The CMOS transistor uses the technique of silicon-on-insulator (SOI), which is very attractive because of the high speed performance, low power consumption, its scalability, and effective potential [3–5]. As compared to the bulk-Silicon substrate, the architecture of SOI MOSFETs is more flexible due to the several parameters such as thicknesses of film and buried oxide, substrate doping as well as back-gate bias which is used for the optimization and scaling. The short channel effects (SCE), junction capacitances, and doping fluctuation are mitigated in ultrathin SOI films [6, 7]. The main advantage of SOI compared to the bulk-Silicon is its compatibility with the use of high resistivity substrates to reduce the substrate coupling and RF losses. However, the numerous advancement for the device architecture have been explored time to time as gate-all-around, delta, lateral epitaxial overgrowth, folded-gate, fin-gate, self-alignment are few of them [8]. The conventional scaling rules suggest that in order to minimize the short channel effects, the doping concentration of the channel must be increased. However, the high doping level degrades the mobility and therefore lowers the drive current. Another possible alternative necessitates the reduction of gate oxide thickness. However, the extent to which gate oxide thickness can be scaled down is limited by the direct tunneling [9–11]. The CSDG MOSFET, which has greater control over the channel, is proposed in order to overcome these drawbacks as well as to offer high packing density and steep subthreshold characteristics. By reducing the thickness of Silicon film of CSDG MOSFET, greater short channel immunity can be achieved. However, as the thickness of the Silicon pillar is reduced, the current drive decreases and thus presenting a serious limiting factor to the device performance.

In ref. [12], the authors have proposed an impressive, compact, and analytical model for the DG MOSFETs, which account for the quantum, volume-inversion [13], short channel, and non-static effects. Electrostatic and Monte-Carlo simulations established the detailed advantage of the DG MOSFET. Djeffal et al. [14] have investigated the scaling capability of the double-gate MOSFET and gate-all-around MOSFETs by using an analytical model of the 2-D Poisson equation in which the hot carrier-induced interface charge effects have been considered. However, based on this analysis, Djeffal et al. [14] have obtained that the degradation becomes more important when the channel length gets shorter and minimum surface potential position is affected by the hot carrier-induced localized interface charge density. By using this analysis, authors [14] also studied the scaling limits of the double-gate MOSFET and gate-all-around MOSFET and compared their performances including the hot carrier effects. As the device size scales down, the total number of channel dopants decreases, which provides a larger variation of dopant numbers significantly affecting the threshold voltage. Dollfus and Retailleau [15] have compared the noise performances of DG MOSFET and SG MOSFETs by a noticeable development of the noise-figure in the double-gate structure that is explored in terms of a favorable increase of cross-correlation between the drain and gate currents. The presence of a residual undesired charged impurity in the channel of a double-gate structure induces perceptible changes in the spectral density of the gate current fluctuations that modifies the noise-figure [16]. Sharma and Kumar [17] have presented a comprehensive analysis to limit the short channel effects in the single-gate MOSFET and double-gate MOSFET. For an RF switch, the bandwidth depends on the capacitance connected to the ground due to the side wall capacitances present in the MOSFET structure [18–21]. These capacitances are also discussed in the following sections for the CSDG MOSFET. Amakawa et al. [22] have developed a surface potential-based model for the cylindrical surrounding MOSFET for which the non-differential equation of the surface potential is not known because of the cylindrical structure.

5.1 Introduction

However, unlike other surrounding-gate MOSFET models, this model includes both the drift and diffusion currents and there is no inherent distinction between the saturation and non-saturation. Amakawa et al. [22] have also demonstrated its accuracy in comparison with device simulation without arbitrary fitting. Li et al. [23] have proposed a novel cylindrical surrounding-gate MOSFETs with electrically induced source/drain extension and demonstrated this with numerical simulation for the first time. In this device, a constant voltage is applied to the side gate to form the inversion layers acting as the extremely shallow virtual source/drain. By using the 3-D device simulator, Li et al. [23] have investigated the device performance focusing on the threshold voltage roll-off, the drain-induced barrier lowering (DIBL), subthreshold swing, and electrical field with carrier temperature. This structure exhibits better suppression of the short channel effects and hot carrier effects when compared to the conventional cylindrical surrounding-gate MOSFETs. Chiang [24] has discussed a model on the basis of quasi-2D potential analysis by using the effective conducting path effect, a concise analytical model for the threshold voltage in the cylindrical fully depleted surrounding-gate MOSFETs. However, besides the increased depth of the effective conducting path, a thin Silicon body and a decreased oxide thickness can reduce threshold voltage roll-off, simultaneously. Kaur et al. [25] have developed a 2-D analytical model for the graded channel fully depleted cylindrical surrounding-gate MOSFET by solving the Poisson's equation in the cylindrical coordinate system. An abrupt transition of the Silicon film doping at the interface has been assumed and the effects of the doping and lengths of the high and low doped regions have been taken into account. The model is used to obtain the expressions of surface potential and electric field in the two regions. The analysis is extended to obtain the expressions for threshold voltage (V_{th}) and subthreshold swing. It is shown that a graded doping profile in the channel leads to suppression of SCEs like threshold voltage roll-off, hot carrier effects, and DIBL. The CSDG MOSFET structure utilizes an undoped body due to the following reasons:

a. Carrier mobility can be enhanced by the undoped body owing to the absence of depletion charges, which can significantly contributes to the effective electric field, thus degrading the mobility [8]
b. The undoped MOSFET can avoid the dopant fluctuation effect, which contributes to the variation of the threshold voltage and drive current. So, this CSDG MOSFET can be used for the purpose of double-pole four-throw (DP4T) RF CMOS switch design [26]

We have proposed a novel model of the low-power and high speed CSDG MOSFET used as RF switch for selecting the data streams from the antennas for both the transmitting and receiving processes. The layout of the proposed design has been studied to understand the effect of device geometry, when working as a switch. However, each parameters of the device are discussed separately for the purpose of clarity of presentation and understanding the operation of CSDG RF CMOS switch.

5.2 Analysis of CSDG RF MOSFET

In a MOSFET, the threshold voltage with the change in source to bulk voltage and body effect is given as [20]:

$$V_{th} = V_{to} + \gamma\left(\sqrt{V_{SB} + 2\phi} - \sqrt{2\phi}\right) \tag{5.1}$$

where V_{th} and V_{to}, are the threshold voltage with substrate bias and the value of threshold voltage at $V_{SB} = 0$, respectively, γ and φ are the body effect parameter and surface potential parameter, respectively. It is notorious that a MOSFET with the nonzero source to body voltage has the threshold voltage modified through the body effect ($V_{SB} \neq 0$). The source-body voltage (V_{SB}) for an n-type MOSFET is $V_B = V_{SS}$ and for a p-type MOSFET $V_B = V_{DD}$, then $|V_{th}|_{V_{SB} \neq 0} \rangle |V_{th}|_{V_{SB}=0}$.

Also, for minimizing the threshold voltage, we have to reduce the body effect parameter to zero, which is a main feature of the CSDG MOSFET. At the turned OFF switch status of a transistor, the current between drain to source should be zero but there is a flow of weak inversion current (subthreshold leakage). In the weak inversion, the drain current varies exponentially with the gate to source bias V_{gs} as given by [21]:

$$I_d \approx I_{d0}\, e^{\frac{V_{gs} - V_{th}}{nV_T}} \tag{5.2}$$

where I_{d0} is the drain current at $V_{gs} = V_{th}$ and $V_T = \frac{k \cdot T}{q}$ is the thermal voltage and $n = 1 + \frac{C_D}{C_{ox}}$ is the slope factor, where C_D and C_{ox} are the capacitance of the depletion layer and capacitance of the oxide layer, respectively. In a long channel device, there is no drain voltage dependence of the current, once $V_{ds} \gg V_{th}$, but as the channel length is reduced, the DIBL introduces drain voltage dependence that depends in a complex way upon the device geometry (channel doping and junction doping). The dopant fluctuations in the source and drain regions overlapped by the gate may strongly influence the series resistance, which should be more important than the effect of a single n-type impurity on the transport in the channel. In general, the effect of diffusion of discrete dopants in the gated region of the Si-film certainly deserves further investigation for such small devices [16, 27].

The basic concept of a CSDG MOSFET is to control the channel very effectively by choosing the channel width to be very small and by applying a gate contact to both sides of the channel, which is of cylindrical type. This concept helps to suppress the SCE and leads to higher currents as compared with a MOSFET having only one gate. It is also observed that an n-doped layer in the channel reduces the threshold voltage and increases the drain current when compared with a device of an undoped channel. The reduction in threshold voltage and increase in the drain current occur with the level of doping. The leakage current is larger than that of an undoped channel but less than that of a uniformly doped channel. In particular, the

5.2 Analysis of CSDG RF MOSFET

CSDG MOSFET with an intrinsic channel is considered as the best candidate for the device downscaling as they offer potential advantages such as:

a. Absence of the dopant fluctuation effect, which contributes to the variation of the threshold voltage and drive current
b. Enhance the carrier mobility owing to the absence of depletion charges which can significantly contribute to the effective electric field, results in degrading mobility

However, the intrinsic channel DG MOSFETs need to rely on gate work-function to achieve multiple threshold voltages on a chip due to the absence of body doping, which is an efficient tool to adjust the threshold voltage. Several efforts including implanted metals fully silicided gates [28–31], alloy metals [32], and metal bilayers [33] have been reported. Although the heavy channel doping can suppress SCEs to some extent and fix the threshold voltage, such high doping will dramatically degrade the mobility in the channel and increase leakage current caused by the band to band tunneling (BTBT) from the channel to the drain and will contribute to threshold voltage variations caused by the significant discrete dopant fluctuation effect [34, 35].

However, doping of the layer with donor atoms can decrease the threshold voltage like full body n-doped DG MOSFETs, but the degradation of the subthreshold swing and increase of the leakage current are less in layer doped DG MOSFETs. Thus, doping a narrow layer rather than doping the entire channel is a better option to tune the threshold voltage while maintaining larger drive current and less leakage current.

We have presented a 2-D physics based modeling of the short channel CSDG MOSFETs of the nanoscale dimensions, also a derivation of precise, self-consistent framework model for the device electrostatics, the drain current, and various capacitances. The modeling has no adjustable parameters and implicitly incorporates scaling with device dimensions and material composition. In the present work, by investigating the influence of source and/or drain circular region SOI MOSFETs, we proposed new design insights to achieve high tolerance to gate misalignment or oversize in the nanoscale devices for low voltage analog and RF applications. Due to the circular source and drain, the gate contact with the source and drain is on a long circular region, which avoids the gate misalignment. The design of CSDG MOSFET shows the following features:

a. Misaligned gate-underlap devices perform significantly better than double-gate devices with abrupt source/drain junctions with identical misalignment
b. Misaligned gate-underlap performance (with source/drain optimization) exceeds perfectly aligned double-gate devices with abrupt source/drain region

The proposed process is an application of DG RF CMOS technology as discussed in Chap. 3, for designing a new CSDG configuration. The choice of the RF CMOS switches as discussed in the Chap. 2, requires an analysis of the performance parameters as the drain saturation current, operating frequency, cut-off frequency, threshold voltage of the n-type MOSFET and p-type MOSFET,

116   5 Cylindrical Surrounding Double-Gate RF MOSFET

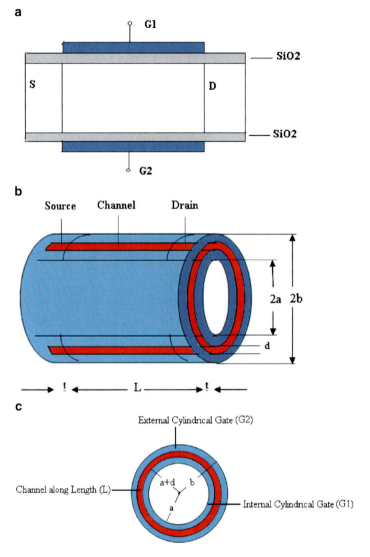

Fig. 5.1 Schematic of (**a**) basic DG MOSFET, (**b**) CSDG MOSFET, and (**c**) cross-section of CSDG MOSFET

control voltage, output power, and forward transconductance [36]. Also this choice requires controlling the increase or decrease of channels for devices, which operates in the depletion region.

Figure 5.1a is a DG MOSFET which has two gates G_1 (front-gate) and G_2 (back-gate). The back-channel has been probed by varying the substrate (back-gate) bias with the front-gate voltage as a parameter. The drain current (I_d) with respect to V_{G_2}

curves is same as front-channel characteristics as strong coupling, linear V_{T_2} (V_{G_1}) variation, constant swing. The effective mobility in the front and back channels is comparable. In the regular SOI MOSFETs, the buried oxide is much thicker (0.4 μm); hence the substrate depletion has a minor effect even on the back-channel. Moreover, the front-channel is hardly affected because it is protected by very small ratio between the thicknesses of the front and back oxides. In the present devices, the substrate effect is exacerbated for several reasons such as the oxide thickness is relatively thin, the front and back oxides have equivalent thickness, and the ultrathin Si-film maximizes the coupling effects [21].

The transconductance decreases more rapidly with the gate voltage in the DG MOSFETs and it indicates a larger value of the mobility degradation factor. Also, the long channel and short channel show the transconductance gain of SG and DG at 300 °K tends to decrease in the short channels. The operation of ultrathin transistors in the CSDG mode brings significant advantages such as scalability, ideal subthreshold slope, high current drive, and excellent transconductance. The gain in transconductance as compared to the SG MOSFET operation based on volume-inversion, which is extremely prominent, has an effect in DG MOSFETs. It does not modify directly the total charge but modifies the carrier profile in the thin film, thus leading to an indirect improvement of the effective mobility.

For the design of CSDG MOSFET, first we design a double-gate MOSFET as shown in Fig. 5.1a, which reveals the n-type DG MOSFET. Similarly we can design the p-type DG MOSFET. Figure. 5.1b is the symmetrical cylindrical surrounding double-gate transistor, which has been presented a compact model that is used for the resistance, capacitance, electrostatic potentials, and current characteristics of the long channel CSDG MOSFETs and cross-section of the CSDG MOSFET is shown in Fig. 5.1c. A mathematical expression of the potential is also derived as a function of the doping concentration in the following sections. After using the expression obtained for the potential and mobile charge, a drain current expression is derived. It is also observed that the threshold voltage shift depends on the doping of substrate and scaling factor. A large scaling factor is preferred to alleviate the threshold voltage degradation [24, 37]. It is noticeable that here two channels are formed, one due to the internal circular gate (G_1) and another due to the outer circular gate (G_2).

5.3 Fabrication Process for CSDG RF MOSFET

For the fabrication of cylindrical surrounding-gate MOSFET, the circular resist dots of different diameters (from 60 to 600 nm) are patterned on the 8 inch. bulk Si-wafer followed by 1 μm deep Silicon etch with SF_6 chemistry under SiN hard mask. Pillars are then oxidized at 1,150 °C to convert into cylindrical form. Then we make a drilling process to make a hollow cylindrical form. The high temperatures are used to decrease the viscosity of the grown oxide, ensuring smooth cylindrical Si-core at the center of the pillar The oxidation rate at the bottom of the pillar is low due to increased stress at high curvature [38, 39] as a result smoothly controlled Silicon footing is formed after oxidation. The gate oxide of ~5 nm was then thermally grown on the exposed wire surface, followed by the deposition of 30 nm poly-Si, which serves as the gate electrode. To reduce the source and drain overlap and physical gate length, the photoresist trimming is used in gate definition. There are three different oxide processes: (1) in situ steam generated oxide with $t_{ox} = 3.3$ nm, (2) rapid thermal oxide with $t_{ox} = 1.5$ nm, and (3) clean oxide annealed in N_2 atmospheres, namely nitride gated oxide with $t_{ox} = 2.8$ nm. The process steps and flows are the same as those originally reported in [40–42]. At last the standard metallization followed to complete the fabrication process of the device.

5.4 Characteristics of CSDG MOSFET

In the conventional bulk MOSFET and SOI devices, the immunity from SCE such as V_{th} roll-off and DIBL requires increasing channel doping to reduce the depletion depth in the substrate. Even when the retrograde channel profiles are used to reduce the mobility degradation and threshold mismatch, this approach intrinsically trades the improved short channel immunity for increased substrate bias sensitivity and degraded subthreshold swing [43]. However, by replacing the substrate with another gate to form a DG MOSFET is shown in Fig. 5.1a, the short channel immunity can be achieved. The threshold voltage (V_{th}) of a fully depleted DG MOSFET can be controlled either by adjusting the channel doping concentration similar to the conventional bulk MOSFET case or by changing the work-function of the gate electrodes. However, in the small gate lengths regime, a CSDG MOSFET is aimed at an undoped Si-channel will be needed to avoid V_{th} fluctuations due to the discrete random dopant placement. If we choose to use an intrinsic Si-channel, then the V_{th} is adjusted by using gate electrodes with mid-gap work-function. In this section, we present a model derived from the 1-D Poisson's equation with all the charge terms included and the channel potential is solved for the asymmetric operation of CSDG MOSFET. Kolberg et al. [44] have presented a model for the short channel DG MOSFETs and found that in the subthreshold regime, the electrostatics of the device is dominated by the capacitive coupling between the electrodes, which is analyzed by conformal mapping techniques. Whereas in the strong inversion regime, the device behavior is dominated by the inversion charge, which allows the 1-D analysis. Djeffal et al. [45] have proposed a new nanoscale graded channel gate-stack DG MOSFET structure and its 2-D analytical model which suppress the SCE and improvement in the subthreshold performances for the nano-electronics applications. This model predicts an incrementing potential barrier (in the surface potential profile along the channel) which ensures a reduced V_{th} roll-off and DIBL effects. Saurabh and Kumar [43] have presented a novel

5.4 Characteristics of CSDG MOSFET

lateral strained double-gate MOSFET. Using device simulation in [43] the authors derived that the strained double-gate MOSFET has a higher ON-current, low leakage, low threshold voltage, excellent subthreshold slope, and better SCEs and also meets important International Technology Roadmap for Semiconductors (ITRS-2010) guidelines [46]. Reyboz et al. [47] have described an explicit compact model of an independent double-gate MOSFET with an undoped channel. This model includes SCE and also mobility reduction, saturation velocity, series resistance, and a charge model. It is applicable for symmetrical, asymmetrical, and independent gate devices. Saad and Ismail [48] have illustrated the process of making a symmetrical self-aligned n-type vertical DG MOSFET over a Silicon pillar. The electron concentration profile is obtained, which demonstrates an increased number of electrons in the channel injected from the source end as the drain voltage increases. The enhanced carrier concentration results to significant reduction in the OFF-state leakage current and improves the DIBL effect and the distinct advantage of the technique reported for suppression of the SCE in nanoscale vertical MOSFET. Dutta et al. [49] have presented a generic surface potential-based V–I characteristics model for the doped and undoped asymmetric DG MOSFET. Taur [27] has led to an analytical solution to the simplified 1-D Poisson's equation in the Cartesian coordinate for a DG MOSFET. However, the interpretation of a surrounding-gate MOSFET can be performed by using the Poisson's equation in the cylindrical coordinate.

The proposed model is able to show the dependency of the front and back surface potential and drain current on the terminal voltages, gate oxide thicknesses, channel doping concentrations, and Silicon body thickness and excellent agreement is observed with the 2-D numerical simulation results. For this purpose, we apply a variable transformation technique to solve the simplified 1-D Poisson's equation in the cylindrical coordinate for the surrounding-gate MOSFET in strong inversion and accumulation region. A general model to describe n-MOSFET in terms of the Poisson's Equation [27]:

$$\nabla^2 \phi = -\frac{q}{\varepsilon_s}\left(-N_a + N_a e^{-q\phi/kT} - \frac{n_i^2}{N_a} e^{q\phi/kT}\right) \quad (5.3)$$

For 1-D situation, the (5.3) can be written as

$$\frac{d^2\phi}{dr^2} + \frac{m}{r}\frac{d\phi}{dr} = -\frac{q}{\varepsilon_s}\left(-N_a + N_a e^{-q\phi/kT} - \frac{n_i^2}{N_a} e^{q\phi/kT}\right) \quad (5.4)$$

where N_a, q, φ, k, and T are the doping concentration Si-substrate, electron charge, Fermi potential, Boltzmann constant, and temperature, respectively. For the double-gate structure in the Cartesian coordinate system, $m = 0$ and for the surrounding-gate structure in the cylindrical coordinate $m = 1$. Also, the boundary conditions for the double cylindrical surroundings structure are

$$\left.\begin{array}{ll}\dfrac{d\phi}{dr}\bigg|_{r=0} = 0 & \text{(a)} \\ \\ \dfrac{d\phi}{dr}\bigg|_{r=a} = \phi_s & \text{(b)} \\ \\ \dfrac{d\phi}{dr}\bigg|_{r=b} = \phi_s & \text{(c)} \end{array}\right\} \quad (5.5)$$

where $r = a$ is the radius of internal cylindrical gate and $r = b$ is the radius of external cylindrical gate as shown in Fig. 5.1c. When the n-type MOSFET is in strong inversion, the holes and depletion charge terms in (5.4) are neglected, and then we get:

$$\frac{d^2\phi}{dr^2} + \frac{m}{r}\frac{d\phi}{dr} = \frac{q}{\varepsilon_s}\frac{n_i^2}{N_a}e^{q\phi/kT} \quad (5.6)$$

We have emphasized on this strong inversion, to study a CSDG MOSFET, the (5.6) with $m = 1$ is written as

$$\frac{d^2\phi}{dr^2} + \frac{1}{r}\frac{d\phi}{dr} = \frac{q}{\varepsilon_s}\frac{n_i^2}{N_a}e^{q\phi/kT} \quad (5.7)$$

The integration technique after solving this with boundary condition of the (5.5), we get the surface charge density:

$$|Q_s| = \left|\frac{d\phi}{dr}\right|_r \varepsilon_s = \frac{kT}{q}\varepsilon_s\sqrt{2B\psi r^2 \exp(q\phi/kT)^s} \quad (5.8)$$

where

$$\psi = \frac{q^2 n_i^2}{\varepsilon_{Si} kT N_a}$$

In the above analysis, the potential and inversion charge concentration both are a function of the distance r from the center of the cylinder for the ultrathin double-gate MOSFET. Here in the proposed device, we have consider the 5 nm thick body (d), $N_a = 10^{20}$ cm^{-3}, $a = 10$ nm, $b = 15$ nm and gate length (L) is 20 nm, which is according to the ITRS-2010 guidelines [46]. Whereas for cylindrical surrounding single-gate a is zero and only b is used. It is expected that the saturation current of a surrounding-gate MOSFET is larger than that of a double-gate structure. In the cylindrical surrounding double-gate, this 5 nm thick body is shared with external

5.4 Characteristics of CSDG MOSFET

circular gate as well as internal circular gate. The surface potential distribution along the channel at

a. $L = 100$ nm
b. $L = 30$ nm.

For $V_{ds} = 0.6$ V, $V_{gs} = 0.1$ V, $t_{si} = 5$ nm, and $t_{ox} = 1.5$ nm are also discussed by the Djeffal et al. [50]. Based on the achieved analytical solutions, we compared the inversion charge concentration of a CSDG MOSFET with that of a CSSG MOSFET [51]. It is found that the significantly higher charge concentration is induced in the surrounding double-gate structure than in the single-gate structure (when the same surface potential is applied), which reveals the better gate control and potentially higher current in the surrounding double-gate MOSFET. The surface potential can be solved in the iteration method if the voltage drop in the gate oxide is considered. The numerical simulation by using Poisson's Equation (5.4) for both the structures has been carried out (as for external gate b is used and for internal gate a is used).

The resistances and capacitance are also present in this layout due to metal connection with output voltage. This drain and source has equal capacitance of 0.19 fF, resistance of 90 Ω, thickness of 2 μm, with metal capacitance of 0.13 fF, diffusion capacitance 0.06 fF, and capacitance of gate is 0.86 fF. The double-gate MOSFET has a resistance of 68 Ω and thickness of 3 μm. Here, we have analyzed the performance of CSDG MOSFET and DG MOSFET with applying gate voltage of 21 μV and pulse width 0.4 ns, rise time and fall time of 50 ps and delay is 1 ps to the capacitive and resistive model of CSDG MOSFET as shown in Figs. 5.3a and 5.4a and the waveform is according to Figs. 5.3b and 5.4b. Assuming that CSDG MOSFET has symmetrical gate structure and voltage applied on both gate (external and internal for cylindrical and upper and lower for simple double-gate) are same. To determine the drain current, a conventional technique in thin-oxide MOSFETs consists of C–V measurements [52, 53]. However, this DG MOSFET consists slightly thick oxide (so that a very small capacitance created); hence the conventional charge $Q = C_{ox} \cdot (V_{gs} - V_{th})$ is suitable, which yields the direct and accurate values for the density of charge carriers, even with double-activated gates. Here C_{ox} is the oxide capacitance and V_{gs} is the gate to source voltage. The linear relationship is complying with $C_{ox} = C_{CSDG} = 2\pi\varepsilon L/\ln(b/a)$. This capacitance $(C_{ox})_{CSDG} > (C_{ox})_{CSSG}$, due to the greater current passing area of the CSDG MOSFET. We find the drain current by using the equation as

$$I_{ds} = \mu \cdot Q \cdot V_{ds} \cdot \frac{W}{L} \tag{5.9}$$

where μ, V_{ds}, W, and L are the channel mobility, applied drain to source voltage, channel width $W = 2\pi(a + b)$ and channel length, respectively. For the CSDG MOSET, $W_{CSDG} = 2\pi(a + b)$ and for CSSG MOSFET, $W_{CSSG} = 2\pi(b)$. So by the (5.9), we have $(I_{ds})_{CSDG} > (I_{ds})_{CSSG}$ as given below:

$$\frac{I_{CSDG}}{I_{CSSG}} = \frac{Q_{CSDG} W_{CSDG}}{Q_{CSSG} W_{CSSG}} \quad \text{or} \quad \frac{I_{CSDG}}{I_{CSSG}} = \frac{C_{CSDG} W_{CSDG}}{C_{CSSG} W_{CSSG}}$$

After replacing the values of capacitances (C_{CSDG} and C_{CSSG}) and widths (W_{CSDG} and W_{CSSG}) in the above, we obtained:

$$\frac{I_{CSDG}}{I_{CSSG}} = \frac{\frac{1}{\ln(b/a)}(a+b)}{\frac{1}{\ln(b)}(b)} \quad \text{or} \quad \frac{I_{CSDG}}{I_{CSSG}} = \frac{\ln(b)}{\ln(b/a)}\left(1+\frac{a}{b}\right) \qquad (5.10)$$

In the CSDG MOSFET, the charge Q is more as compared to the DG MOSFET due to the higher capacitance values, so the drain current is higher in the CSDG MOSFET devices as compared to the simple DG MOSFET and the CSSG MOSFET. When the metal-gate work-function is raised, the leakage current (I_{OFF}) decreases extensively and threshold voltage increases [24]. In order to maintain the I_{OFF} very low for a switch, it is necessary to increase the metal work-function or resistance at OFF-state (R_{OFF}). It should be very high as well as resistance at ON-state (R_{ON}) should be low (as shown in the Sect. 5.4). Moreover, the increase in metal work-function is accompanied with an increase in the threshold voltage. The output voltage can be achieved after obtaining the threshold voltage. So, we conclude from these parameters that output voltage stabilization for CSDG MOSFET is less as compare to SG MOSFET. Also, if we increase the gate lengths, then C_{ox} will increase and drain to source current I_{ds} reduces.

5.5 Resistive and Capacitive Model of the CSDG MOSFET

In this section, we have presented a precise two-dimensional resistive and capacitive model for the nanoscale CSDG MOSFETs as shown in Fig. 5.2 covering a wide range of the operating regions, geometries, and material combinations. The electrical resistance of an object measures its opposition to the passage of an electric

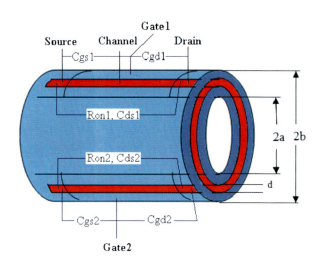

Fig. 5.2 Model of CSDG MOSFET transistor with its components at ON-state

5.5 Resistive and Capacitive Model of the CSDG MOSFET

current. An object of uniform cross-section has a resistance proportional to its resistivity and length as well as inversely proportional to its cross-sectional area. For the given design of CSDG MOSFET, under the operating condition, the insertion loss is conquered by its ON-state resistance $R_{CS} = \rho L/A$ and substrate resistance, where ρ, L, and A are the resistivity, length of channel, and cross-section area in which the current flows, respectively.

Now, this proposed cylindrical surrounding structure has following two types of resistances:

a. Due to the current flow in channel-1 with respect to the internal gate, $R_{CS\ 1} = \rho L/\pi((a+t)-(a))^2$ which becomes $R_{CS\ 1} = \rho L/\pi(t)^2$
b. Due to the current flow in channel-2 with respect to the external gate, $R_{CS\ 2} = \rho L/\pi((b)-(b-t))^2$ which becomes $R_{CS\ 1} = \rho L/\pi(t)^2$

where t is the junction depth (thickness of source and drain, $L > t$). An effective area is that area in which current flows is $\pi(b-a)^2$ and channel length is L. So the effective ON-resistance for this architecture is

$$R_{\text{ON-CSDG}} = \frac{\rho L}{\pi(b-a)^2} \tag{5.11a}$$

Now, for the CSSG MOSFET, an internal gate-1 is not present so the only external gate-2 is responsible for the ON-state resistance. The total maximum ON-state resistance across source to drain for the CSSG is [3]

$$R_{\text{ON-CSSG}} = \frac{\rho L}{\pi(b)^2} \tag{5.11b}$$

$$\frac{R_{\text{ON-CSDG}}}{R_{\text{ON-CSSG}}} = \frac{1/(b-a)^2}{1/(b)^2} = \left(\frac{b}{b-a}\right)^2 \tag{5.12}$$

The capacitive model of a CSDG MOSFET transistor which is biased in the linear region, at the ON-state is shown in Fig. 5.3a. Here the isolation of this MOSFET for the switching application is finite due to signal coupling through the parasitic and junction capacitances. So, the capacitances for the CSDG MOSFET design will follow the cylindrical coordinate as

$$C_{\text{CSDG}} = \frac{2\pi\varepsilon L}{\ln\left(\frac{b}{a}\right)} \tag{5.13}$$

where ε, a, b, and L are the dielectric permittivity, inner radius, outer radius, and length of cylinder, respectively.

This cylindrical surrounding structure has six types of capacitances as given by (5.14a–5.14f), where d is the depth of source and drain from the surface of gates toward the substrate thickness ($t \sim d$). All these six capacitances are shown

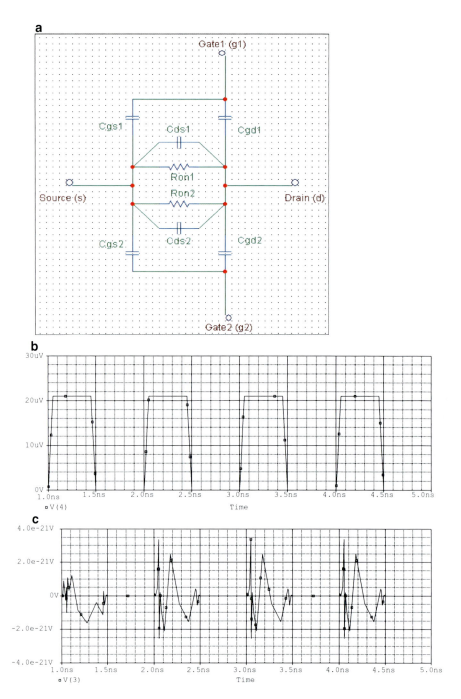

Fig. 5.3 Design of the CSDG MOSFET with SPICE (**a**) capacitive models operating as a switch at ON-state, (**b**) input signal applied to gates, (**c**) output signal at drain, (**d**) source current variation with frequency, and (**e**) drain current variation with frequency

5.5 Resistive and Capacitive Model of the CSDG MOSFET 125

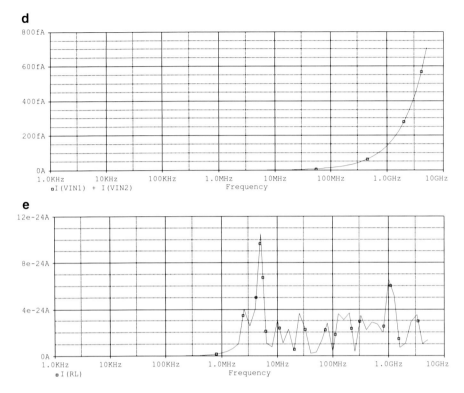

Fig. 5.3 (continued)

in Fig. 5.3a. For the CSDG MOSFET when both the transistors are ON, then the capacitances due to bulk with source and drain as C_{sb} and C_{db} are not present because the substrate is not present in CSDG MOSFET structure. So fewer signals being coupled to the substrate and also no dissipation into the substrate resistance R_b. When the transistor is in the cut-off region, increasing the C_{ds1}, C_{ds2}, C_{gd1}, C_{gd2}, C_{gs1}, and C_{gs2} leads to higher isolation between the source and drain due to no capacitive coupling between these terminals. Whereas for the single-gate MOSFET, when the transistor is ON, increasing C_{sb} and C_{db} leads to more signal being coupled to the bulk and dissipated in the bulk resistance R_b. At the transistors cut-off region C_{ds}, C_{gd}, and C_{gs} increase which directs to lower isolation between the source and drain due to capacitive coupling between these terminals.

After designing the CSDG MOSFET, we have simulated this by using the Cadence's Simulation Program with Integrated Circuit Emphasis (SPICE), a general purpose open source analog electronic circuit simulator (Fig. 5.3) and Agilent's Advanced Design Simulator (ADS), an EDA tool (Fig. 5.4) for 10 MHz to 5.0 GHz. ADS is an electronic design automation software for the RF, microwave, and other high-speed circuit applications, which provides full standard-based design

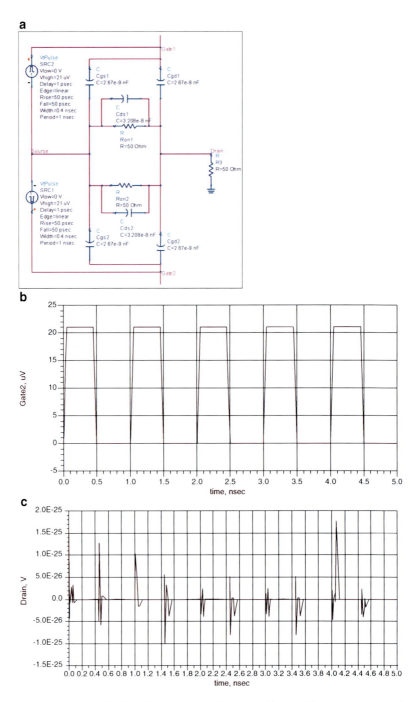

Fig. 5.4 Design of the CSDG MOSFET with ADS (**a**) capacitive models operating as a switch at ON-state, (**b**) input signal applied to both gates, and (**c**) output signal at drain

5.5 Resistive and Capacitive Model of the CSDG MOSFET

and verification with wireless libraries and circuit system electromagnetic co-simulation in an integrated platform.

1. Between gate to source

$$\text{Due to internal gate} \quad C_{gs1} = \frac{2\pi\varepsilon t}{\ln\left(\frac{a+d}{a}\right)} \qquad (5.14\text{a})$$

$$\text{Due to external gate} \quad C_{gs2} = \frac{2\pi\varepsilon t}{\ln\left(\frac{b}{b-d}\right)} \qquad (5.14\text{b})$$

2. Between gate to drain

$$\text{Due to internal gate} \quad C_{gd1} = \frac{2\pi\varepsilon t}{\ln\left(\frac{a+d}{a}\right)} \qquad (5.14\text{c})$$

$$\text{Due to external gate} \quad C_{gd2} = \frac{2\pi\varepsilon t}{\ln\left(\frac{b}{b-d}\right)} \qquad (5.14\text{d})$$

3. Between drain to source

$$\text{Due to internal gate} \quad C_{ds1} = \frac{2\pi\varepsilon L}{\ln\left(\frac{a+d}{a}\right)} \qquad (5.14\text{e})$$

$$\text{Due to external gate} \quad C_{ds2} = \frac{2\pi\varepsilon L}{\ln\left(\frac{b}{b-d}\right)} \qquad (5.14\text{f})$$

According to Fig. 5.3a and from (5.14a–5.14f), we have computed the $C_{gs1} = C_{gs2} = 2.67 \times 10^{-9}$ nF, $C_{gd\,1} = C_{gd\,2} = 2.67 \times 10^{-9}$ nF, and $C_{ds\,1} = C_{ds\,2} = 3.208 \times 10^{-8}$ nF. Here, we have applied the input voltage of 21 μV, with pulse width 0.4 ns, rise time and fall time of 50 ps each and delay is 1 ps. In both of the simulation results, we obtained that the drain current flows only when voltage supply is ON and no current flows when the voltage supply is OFF. It means switch has a clear cut OFF-status, which is a good sign of switching isolation in the CSDG MOSFET.

In Figs. 5.3a and 5.4a, for the CSDG MOSFET, the total capacitance across source to drain is

$$C_{CSDG} = C_{ds1} + C_{ds2} + \frac{C_{gs1}.C_{gd1}}{C_{gs1} + C_{gd1}} + \frac{C_{gs\,2}.C_{gd\,2}}{C_{gs\,2} + C_{gd\,2}} \qquad (5.15\text{a})$$

Now, for the CSSG MOSFET, internal gate-1 is not present. So, only the external gate-2 is responsible for the capacitances. The total capacitance across source to drain is [3]

$$C_{CSSG} = C_{ds2} + \frac{C_{gs2}.(C_{gd2} + C_{gb2})}{C_{gs2} + C_{gd2} + C_{gb2}} + \frac{C_{sb2}.C_{db2}}{C_{sb2} + C_{db2}} \qquad (5.15\text{b})$$

where C_{gb} is capacitance from gate to bulk connections. For an example, if each capacitance is of 1 pf, then $C_{CSDG} = 1.4 C_{CSSG}$. With the application of these capacitances and supply voltage, we can calculate the energy stored with the proposed device as

$$\left. \begin{array}{ll} U_{CSDG} = \dfrac{C_{CSDG} V^2}{2} & (a) \\\\ U_{CSSG} = \dfrac{C_{CSDG} V^2}{2} & (b) \end{array} \right\} \quad (5.16)$$

So, the ratio of these two stored energies will become

$$\frac{U_{CSDG}}{U_{CSSG}} = \frac{C_{CSDG}}{C_{CSSG}} \quad (5.17)$$

With the (5.17) at 1 pf of capacitances, the $U_{CSDG} = 1.4\, U_{CSSG}$. Therefore, the CSDG has more energy. Since by the calculation of capacitances with the (5.15a) and (5.15b), we obtained that the capacitance $C_{CSDG} > C_{DG} > C_{SG}$ that reveals the isolation is better in the CSDG compared to the simple DG MOSFET and SG MOSFET. Also, the resistance $R_{CSDG} < R_{DG} < R_{SG}$, which shows that the current flow from source to drain in the CSDG MOSFET, is better than the DG MOSFET and SG MOSFET. For appropriate working of a switch and to reduce the insertion loss, we can also achieve further reduction in ON-state resistance with choosing large mobility (μ), increased aspect ratio (W/L), and keeping $V_{gs} - V_{th}$ large as it is clear from the (5.9) [50].

Figure 5.3c justified the output signal achieved when the input signal is applied on both the gate according to Fig. 5.3b. This output signal is same for all the positive edge duration (0.5 ns) and current increases exponentially as shown in Fig. 5.3d, and a very small drain current flow for that duration as shown in Fig. 5.3e. Similarly, Fig. 5.4c shows the excitation when input signal changes from low to high (0 V to 21 µV) or high to low (21 µV to 0 V) as shown in Fig. 5.4b. After the calculation of the above parameters from the (5.9) to (5.17), we have summarized these results as shown in Table 5.1.

Table 5.1 reveals the comparison of CSDG MOSFET with CSSG MOSFET parameters, which is closed to the aspect of RF CMOS switch design. Key figures of merit of a transceiver switch are the insertion loss and power handling capability measured by the power 1 dB compression point. The ON-state resistance of the transistor is one of the dominant factors to determining insertion loss. The drain-to-body and source-to-body junction capacitances-associated parasitic resistances due to the conductive nature of Silicon substrates are also critical factors to determining insertion loss [54].

Since the insertion loss is proportional to $(R_{ON})^2$, therefore for the proposed design, the insertion loss becomes one-fourth of the existing models as the capacitances present in the CSDG MOSFET device is 1.4 times than the CSSG

5.6 Explicit Model of CSDG MOSFET

Table 5.1 Comparison of the various circuit parameters of the CSDG MOSFET and existing CSSG MOSFET model

Parameters	CSDG MOSFET	CSSG MOSFET
Gate/control voltage	1.2 V	1.2 V
Total capacitance (S to D)	1.4C_{SG}	C_{SG}
ON-resistance (R_{ON})	0.5R_{ON}	R_{ON}
Thickness of oxide layer	3 μm	2 μm
Resistance of Poly/Gate	68 Ω	32 Ω
No. of capacitors	6	6
Bulk capacitor	No	Yes
Gain (=1 upto)	0.60 V	0.40 V
Energy stored	1.4U_{SG}	U_{SG}

MOSFET device as well as ON-resistance of CSDG MOSFET is half of the CSSG MOSFET. These properties make the CSDG MOSFET design useful for application in the DP4T RF CMOS switch. Due to thin-oxide layer of CSDG MOSFET, this device can be fabricated in a small size of chip as compared to the CSSG MOSFET. We also conclude that the gain is constant up to 0.60 V for CSDG MOSFET compared to 0.40 V of CSSG MOSFET.

5.6 Explicit Model of CSDG MOSFET

Assuming the gradual channel approximation in an undoped (lightly doped) n-type CSDG MOSFET as shown in Fig. 5.2. The Poisson's Equation can be written as

$$\frac{d^2\psi}{dr^2} + \frac{1}{r}\frac{d\psi}{dr} = \frac{qn_i}{\varepsilon_{si}} e^{q(\psi-V)/kT} \tag{5.18}$$

where q, n_i, and ε_{Si} are the electronic charge, intrinsic carrier concentration, and dielectric permittivity of Silicon, respectively. V and ψ are the electrostatic potential and electron quasi-Fermi potential, respectively. It has been assumed that the hole density is negligible as compared with the electron density. Equation 5.18 satisfies the following boundary conditions:

$$\frac{d\psi}{dr} = \begin{cases} 0 & r = 0 \\ \psi_{\text{int-surface}} & r = a \\ \psi_{\text{ext-surface}} & r = b \end{cases} \tag{5.19}$$

where $\psi_{\text{int-surface}}$ stands for the internal gate surface and $\psi_{\text{ext-surface}}$ stands for the external gate surface potential. The current mainly flows along the direction of channel for both the gates. Therefore, we can assume that is constant along the direction. Equation (5.18) can be solved analytically as [55, 56]

$$\psi(r) = V + \frac{kT}{q} \log\left(\frac{-8AkT\varepsilon_{Si}}{q^2 n_i (1 + Ar^2)^2}\right) \quad (5.20)$$

where A is related to ψ_S through the second boundary condition in (5.19). The total mobile charge (per unit gate area) can be written as

$$Q = C_{ox}\left[V_{gs} - \Delta j - (\psi_{\text{int-surface}} + \psi_{\text{ext-surface}})\right] \quad (5.21)$$

where $C_{ox} = \varepsilon_{ox}/R\ln\left(1 + \frac{t_{ox}}{R}\right)$ and $\Delta\varphi$ is the work-function difference between the gate electrode and intrinsic silicon. From the Gauss's law, the following relation holds:

$$Q = C_{ox}\left[V_{gs} - \Delta j - (\psi_{\text{int-surface}} + \psi_{\text{ext-surface}})\right] = \varepsilon_{Si}\frac{d\psi}{dr}\bigg|_{r=a} + \varepsilon_{Si}\frac{d\psi}{dr}\bigg|_{r=b} \quad (5.22)$$

By substituting (5.20) into (5.22), we get the following relation where r equal to a internal radius or b is external radius:

$$\frac{q(V_{gs} - \Delta j - V)}{kT} - \log\left(\frac{8kT\varepsilon_{Si}}{q^2 n_i r^2}\right) + \log\left(\frac{(1 + Ar^2)^2}{Ar^2}\right) + \frac{Ar^2}{1 + Ar^2} = 0 \quad (5.23)$$

For a given V_{gs}, A can be solved from (5.23) as a function of V. Here V varies from the source-to-drain, being $V = 0$ at the source end and $V = V_{ds}$ at the drain end. From this analysis, we can obtain a charge control model relating to the carrier charge density with the bias. The drain current in terms of the carrier charge densities is calculated from

$$I_{ds\text{-int}} = \mu \frac{2\pi a}{L} \int_0^{V_{ds}} Q(V) dV \quad (5.24a)$$

$$I_{ds\text{-ext}} = \mu \frac{2\pi b}{L} \int_0^{V_{ds}} Q(V) dV \quad (5.24b)$$

In the above discussion, we realized that the current into the external cylindrical MOSFET is greater than the current into internal cylindrical MOSFET. However, both of these current gives the overall drain current for the CSDG MOSFET.

5.7 Gate Leakage Current, Noise Model, and Short Channel Effects for CSDG MOSFET

The noise is unwanted electrical or electromagnetic energy that degrades the quality of signals and data. In RF circuit designs, the noise is an important issue to be considered. However, both passive and active components in the circuit generate various types of noise. The circuit noise is generated by the electrical components such as resistors and transistors. The noise in RF design appears as either additive noise or phase noise. The additive noise is the noise changing the amplitude of the required signal, while phase noise is the noise changing the phase of the wanted signal. To understand the noise behavior, a single MOSFET can be considered as a small circuit with different resistive, capacitive, and active components as we have seen in the previous chapter. Thus different noise sources exist in a MOS transistor with their power spectral densities. However, such as terminal resistance thermal noise at the gate, drain and source, flicker noise in the channel, substrate resistance thermal noise and induced gate noise.

The Flicker noise or pink noise or $1/f$ noise is caused by the random trapping of the charge at the oxide–silicon interface of MOS transistors. It is a low frequency noise and it mainly affects the LF performance of the device, so it can be ignored at very high frequency (VHF), but the contribution of flicker noise should be considered in designing RF circuits such as mixers, oscillators, or frequency dividers that up-convert the LF noise to higher frequency and deteriorate the phase noise or the signal to noise ratio. The channel resistance and all terminal resistances contribute to the thermal noise at HF, but the channel resistance dominates in the contributions of the thermal noise from the resistances in the device. The induced gate noise is generated by the capacitive coupling of local noise sources within the channel to gate.

In the CSDG MOSFET devices with ultrathin gate oxide, direct tunneling is a dominant mechanism of gate leakage current. This current can be divided into six major (three due to internal gate and three due to external gate) contributions [57, 58]: the gate to inverted channel current (I_{gc1} and I_{gc2}), the gate to source (I_{gs1} and $I_{gs\,2}$) and the gate to drain (I_{gd1} and I_{gd2}) components due to the path through the source and drain overlap regions. The gate leakage current noise performances of a CMOS device can be characterized in terms of the gate noise current spectrum, which can be modeled by [59]

$$S^2 = i^2 + N^2$$
$$= 2qI\Delta f + \frac{k}{f^n}\Delta f \qquad (5.25)$$

where i^2 and N^2 describe the shot noise current and Flicker noise current. The q, I, and Δf are electronic charge, gate current, and noise bandwidth, respectively. k is the empirical parameter that is device specific and n is an exponent that is usually close to unity. The term I in (5.25) can be expressed by means of the shot noise law $I = I_{g1} + I_{g2}$, where I_{g1} and I_{g2} are the sum of the absolute values of each gate current contribution for a given bias condition [60]. The shot noise occurs in conducting p–n junctions. The shot noise current depends on the charge of electron, the total DC current flow, and the bandwidth. However, the thermal noise is caused by the random thermally generated motion of electrons. It depends on the absolute temperature. The thermal noise related to the substrate resistance (R_{sub}) can produce assessable effects at the terminals of the device. The thermal noise produced by this substrate resistance modulates the potential of the back-gate contributing some noisy drain current of a MOSFET is given by

$$i^2_{nd, sub} = 4kTR_{sub}g^2_{mb}\Delta f \qquad (5.26)$$

However, for the CSDG MOSFET, bulk/substrate is not present, so $R_{sub} = 0$ which provides $i_{nd,sub}^2 = 0$. Hence no noise is produced by the substrate resistance. In addition to the drain current noise, the thermal agitation of the channel charge has another important consequence as gate noise. The fluctuating channel potential couples capacitive into the gate terminal, leading to a noisy gate current. The noisy gate current may also be produced by thermally noisy resistive gate material. Although, this noise is negligible at low frequencies, however it can be dominate at radio frequencies.

Here we have presented an analytical model for undoped CSDG MOSFETs and provide the explicit solutions for the intermediate parameters that is used in DG MOSFETs models, which has been validated by numerical simulations. The proposed model is based on the charge control process by which we derived a channel current expression in terms of the charge densities available in the channel at the source and drain ends of the device. This model can be an explicit model if we use the suitable expressions for the charge densities with the applied voltages. The channel charge distribution in the silicon film is sufficiently accounted for charge control model. The channel current expression presents an infinite order of continuity over all operating regions, which formulate the model capable for circuit simulation. In this proposed model the terminal charges, drain impedance, and transconductance can be expressed as explicit functions of applied voltages and various structural parameters.

5.8 Cross talk in CSDG MOSFET Model

In ref. [61], the authors have analyzed the substrate cross talk into high resistive Silicon substrate and discussed the impact on the RF behavior of SOI MOSFET. The introduction of high resistivity Si-substrate has converted Silicon into a suitable technology for high frequency applications. However, it is known that oxidized high resistivity substrate suffers from parasitic surface conduction effect. The positive fixed charges inside the oxide attract electrons to the interface and creating an inversion and accumulation layer at the Si–SiO$_2$ interface. This thin highly conductive layer is responsible for the substrate losses. This issue can be overcome by introducing a trap-rich passivation layer between the oxide and the

5.8 Cross talk in CSDG MOSFET Model

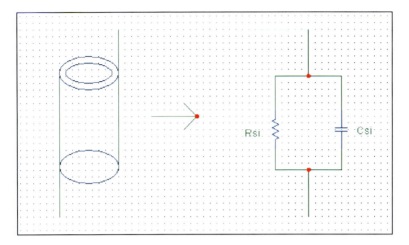

Fig. 5.5 Equivalent resistive and capacitive model of the CSDG MOSFET

high resistivity Si-substrate, which captures the free carriers and locally depletes the high resistivity Si-substrate. Several techniques have been used to generate such trap-rich layer as micro-machined structures [62], ion implantation, and deposition of amorphous Silicon or polycrystalline Silicon layer, which is easily suitable for the CSDG MOSFET structure. Silicon, as any semiconductor material, exhibits both conductive (resistive effect) and dielectric (capacitive effect) characteristics.

At the frequencies below a certain crossover (f_s) the conductive nature of the semiconductor dominates over the dielectric behavior. Thus, the substrate can be modeled as purely resistive. The conductivity for a doped semiconductor is given as $\sigma = q(p\mu_p + n\mu_n)$, where q, μ_n, and μ_p are the electron charge, mobility of the electrons and holes carriers, respectively, and n and p stand for respective carrier densities. While the effective carrier mobility depends also on the number of carriers (scattering effect), the above expression is dominated by the carrier concentration means that the conductivity is an increasing function of the carrier densities.

At the frequencies above the crossover, the dielectric behavior of the semiconductor cannot be neglected, thus the substrate must be modeled as a resistive and capacitive network as shown in Fig. 5.5. In the frequency domain the equivalent admittance Y_{si} for a piece of substrate is given by

$$Y_{si} = \frac{1}{R_{Si}} + j\omega C_{Si} \tag{5.27}$$

and

$$T_{si} = R_{Si} C_{Si} \tag{5.28}$$

where R_{Si}, C_{Si}, ω, and T_{Si} are the substrate equivalent resistance, capacitance, angular velocity, and the time constant. For the cylindrical surrounding structure the resistance and capacitance are as

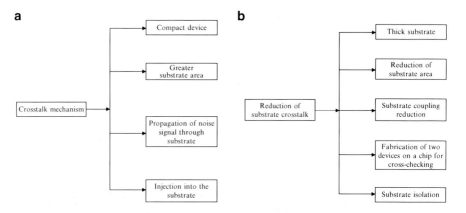

Fig. 5.6 (a) Substrate cross talk mechanism and (b) Reduction of cross talk with CSDG MOSFET model

$$R_{CSDG} = \frac{\rho_{Si} dL}{dA} \tag{5.29}$$

and

$$C_{CSDG} = \frac{\varepsilon_0 \varepsilon_{Si} dA}{dL} \tag{5.30}$$

where ε_0, ε_{Si}, dL, and dA are the vacuum dielectric permittivity, Silicon permittivity, elementary length of the small piece of homogenous substrate and its elementary area $\pi(b-a)^2$. So we obtained the crossover frequency as

$$f_T = \frac{1}{2\pi T_{Si}} = \frac{Q(p\mu_p + n\mu_n)}{2\pi \varepsilon_0 \varepsilon_{Si}} \tag{5.31}$$

which is independent from dimensions of conducting path available in the CSDG MOSFET. As it can be seen in the (5.31), that the T_{Si} is not related to the dimensions of the considered semiconductor volume but only to the substrate electrical properties. To operate the device at lower frequency, the vacuum dielectric permittivity and Silicon permittivity should be accordingly higher. At low frequencies, the substrate resistance, R_{Si}, is more important and the associated capacitance, C_{Si}, can be neglected. As the pulsation ω increases, the impedance relative to the capacitive effect decreases to become equal to that of the resistive effect at the crossover frequency (f_T) defined by the (5.31). Figure 5.6a shows the mechanism of the cross talk in a substrate. To overcome with the problem of substrate cross talk, we include following process in proposed CSDG MOSFET as given in Fig. 5.6b.

5.9 Advantages of the CSDG MOSFET Model

Gidon [63] has used the MOS transistor model from COMSOL (a multiphysics modeling and simulation software) as a template to propose the model of DG MOSFET for the purpose of resolving the SCE problems in the MOSFET configurations. However, such configurations are directly related to the continuous reduction of the device size in the microelectronic technology. The drain current versus gate voltages is considered as parameters. Tamer and Roy [64] have discussed the DG MOSFET structures such as back-gate, metal-gate work-function engineering, and gate isolation processes offer attractive options for circuit design in multi-gate transistor. Inserting extra independent gate processes into fabrication flow allows the exploitation of undoped ultrathin body-associated strong gate-to-gate coupling in DG MOSFET structures. In [64], the authors also demonstrated that under a leakage constraint, the DG MOSFET circuits provide the best power performance trade-off with symmetrical devices. Li and Chou [65] have proposed a unified 2-D density gradient model, which has simulated 10 nm DG MOSFETs and obtained the DG MOSFETs with thinner Si-films, which significantly suppress the SCE, but the ON-state current issue suffers. A compromise between Si-film thickness and gate channel length has to be maintained at the same time such that an optimal device characteristic could be achieved. Razavi and Orouji [66] proposed a model to reduce the SCE of nanoscale DG MOSFET and to improving the reliability of the device, a triple material DG MOSFET is used with different workfunctions. However, based on this simulation results, it reveals that the DG MOSFET exhibits reduced SCEs such as DIBL, hot carrier effect, and improved reliability. Also, it described that the DG MOSFET leads to simultaneous enhancement of the transconductance and reduction of the drain conductance, which itself leads to higher DC gain in comparison with the conventional DG MOSFET. The better reduction of the SCE and improvement of the device reliability could be expected by changing the length ratio of the gate materials with optimization. Subramanian [67] has discussed the multi-gate FET for which author admits about the problem of high paracitics and fin roughness.

Kim et al. [68] proposed a Surrounding-Gate MOSFET with Vertical Channel (SGVC) cells as a 1-T DRAM cell. To confirm the memory operation of the SGVC cell, Kim et al. have simulated its memory effect and fabricated the highly scalable SGVC cell. According to the simulation and measurement results, the SGVC cells can operate as a 1-T DRAM having a sufficiently large sensing margin. Venugopalan et al. [69] have developed a Berkeley short channel insulated-gate FET model for compact model of cylindrical/surround gate MOSFET (BSIM-CG) for circuit simulations which is a production circuit simulation ready compact model for cylindrical/surround gate transistors. The advantage of aforementioned references with models for CSSG MOSFET is comparable with the proposed CSDG MOSFET, which is summarized in Table 5.2.

Table 5.2 Advantage of the proposed CSDG MOSFET model over several reported literatures for CSSG MOSFET

S. No.	Properties	Reported literature	CSDG MOSFET
1	Effective conducting path	[24] Designed a concise analytical model for the threshold voltage in cylindrical surrounding-gate MOSFETs. Besides the increased depth of the effective conducting path, a thin silicon body and a decreased oxide thickness can reduce threshold voltage	Increased depth of the effective conducting path (due to two surroundings), a thin Silicon body and a decreased oxide thickness which reduces threshold voltage
2	Structure	[37] Presented a structure of cylindrical surrounding gate with the rotation of single-gate MOSFET along the substrate region	This structure is designed by the rotation of double-gate MOSFET along any one gate
3	Cross-section	[40] Described the ohmic source/drain contacts, ohmic shape, triangular, and circular cross-sections, which have been demonstrated with a minimum Silicon circular diameter of 5 nm	This structure has only circular and cylindrical cross-sections which have been demonstrated with a minimum circular diameter of 5 nm to avoid Quantum effects
4	Compact model	[43] Presented a double-gate TFET (SDGTFET). Using device simulation proved that the SDGTFET has a higher ON-current, low leakage, low threshold voltage, excellent subthreshold slope, and good short channel effects	Compact model also satisfied with CSDG MOSFET as an independent DG MOSFET with an undoped channel which also includes SCE and ON-resistance and a charge model, for symmetrical, asymmetrical, and independent gate devices
5	Fabrication process	[47] Revealed a process of making a symmetrical self-aligned n-type vertical double-gate MOSFET over a Silicon pillar	CSDG MOSFET is similar to process of making a symmetrical self-aligned n-type vertical double-gate MOSFET over a hollow Silicon pillar
6	Saturation region	[63] For the purpose of resolving the SCE problems in MOSFET structures and to avoid being in the saturation region, this model has such architectures that directly related to the constant reduction of the feature size of device. The drain current versus gate voltages considered as parameters	CSDG MOSFET easily avoids being in the saturation region

(continued)

5.10 Conclusions

Table 5.2 (continued)

S. No.	Properties	Reported literature	CSDG MOSFET
7	Replacement of multi-gate process	[64] Inserted some extra independent gates process into fabrication flow to allow the exploitation of undoped ultrathin body-associated strong gate-to-gate coupling in DG MOSFET	It is a replacement of independent multi-gate process
8	Si-film thickness and channel length	[65] Compromised between Silicon film thickness and gate channel length should be maintained so that an optimal device characteristic could be obtained	Silicon film thickness and gate channel length is maintained
9	Problem of high parasitic	[67] Discussed the multi-gate FET, for which he admits about the problem of high paracitics and fin roughness	This design has overcome the problem of high paracitics and fin roughness
10	1-T DRAM cell	[68] Proposed a surrounding-gate MOSFET with vertical channel (SGVC) cell as a 1-T DRAM cell. According to simulation and measurement results, the SGVC cell can operate as a 1 T DRAM having a sufficiently large sensing margin	Two 1-T DRAM cell can be fabricated with the CSDG MOSFET on a single chip
11	Material	[70] It can be seen that the DG MOSFET leads to simultaneous enhancement of transconductance and reduction of drain conductance which itself leads to higher DC gain in compare with the conventional DG MOSFET. Better reduction of the SCE and improvement of the device reliability could be expected by changing the length ratio of the gate materials and optimizing them	This facility is achieved with only one material CSDG MOSFET

5.10 Conclusions

In this chapter, we have designed and simulated a CSDG MOSFET for the parameters available in this design by using the PSPICE and ADS. It includes the basics of the circuit elements parameter required for the radio frequency subsystems of the integrated circuits such as drain current, output voltage, threshold voltage, capacitances, resistances at switch ON-state condition, oxide thickness, resistance of polysilicon, energy stored, cross talk, number of bulk capacitors, and power or voltage gain. For the RF/microwave switch, we achieved the process to minimize the control voltage, capacitances for isolation, and the resistance for the switching condition and increased energy storage of a device.

However, the mobile charge density is calculated by using the analytical expressions obtained from modeling the surface potential as well as the difference of potentials at the surface and at the center of the Silicon-doped layer without solving any transcendental equations. The analytical expressions for the charge characteristics are presented as the function of Silicon layer impurity concentration, gate dielectric, and Silicon layer thickness with the variable mobility. The transfer characteristics in linear and saturation regions, as well as for the output characteristics verify the good agreement within the practical range of gate and drain voltages, as well as gate dielectric and Silicon layer thicknesses. In this design, the transistor width can be increased for CSDG MOSFET, therefore the peak power-added efficiency and output power decreases due to a reduction in the maximum frequency.

References

1. S. Ahmed, C. Ringhofer, and D. Vasileska, "An effective potential approach to modeling 25 nm MOSFET devices," *J. of Computational Electronics*, vol. 9, no. 3–4, pp. 197–200, Oct. 2010.
2. Viranjay M. Srivastava, K. S. Yadav, and G. Singh, "Analysis of drain current and switching speed for SPDT switch and DPDT switch with the proposed DP4T RF CMOS switch," *J. of Circuits, Systems and Computers*, vol. 21, no. 4, pp. 1–18, June 2012.
3. Viranjay M. Srivastava, K. S. Yadav, and G. Singh, "Design and performance analysis of double-gate MOSFET over single-gate MOSFET for RF Switch," *Microelectronics Journal*, vol. 42, no. 3, pp. 527–534, March 2011.
4. International Technology Roadmap for Semiconductors-2010, www.public.itrs.net
5. S. Cristoloveanu and S. S. Li, *Electrical Characterization of SOI Materials and Devices*, Kluwer Publications, Massachusetts, USA, 1995.
6. N. Ashraf and D. Vasileska, "1/f Noise: threshold voltage and ON-current fluctuations in 45-nm device technology due to charged random traps," *J. of Computational Electronics*, vol. 9, no. 3–4, pp. 128–134, Oct. 2010.
7. K. Takeuchi, T. Fukai, A. Nishida, and T. Hiramoto, "Understanding random threshold voltage fluctuation by comparing multiple fabs and technologies," *Proc. of IEEE Int. Electron Device Meeting*, Washington, DC, USA, 10–12 Dec. 2007, pp. 467–470.
8. Yang Tang, Li Zhang, and Yan Wang, "Accurate small signal modeling and extraction of silicon MOSFET for RF IC application," *Solid State Electronics*, vol. 54, no. 11, pp. 1312–1318, Nov. 2010.
9. R. H. Yan, A. Ourmazd, and K. F. Lee, "Scaling the Si MOSFET: from bulk to SOI to bulk," *IEEE Trans. on Electron Devices*, vol. 39, no. 7, pp. 1704–1710, July 1992.
10. A. Nitayami, H. Takato, N. Okabe, K. Sunouchi, K. Hiea, and F. Horiguchi, "Multipillar surrounding gate transistor (M-SGT) for compact and high-speed circuits," *IEEE Trans. on Electron Devices*, vol. 38, no. 3, pp. 579–583, March 1991.

References

11. S. Watanabe, K. Tsuchida, D. Takashima, Y. Oowaki, A. Nitayama, and K. Hieda, "A novel circuit technology with surrounding gate transistors (SGT's) for ultra high density DRAM's," *IEEE J. Solid State Circuits*, vol. 30, no. 9, pp. 960–971, Sept. 1995.
12. L. Ge and Jerry G. Fossum, "Analytical modeling of quantization and volume inversion in thin Si film DG MOSFETs," *IEEE Trans. on Electron Devices*, vol. 49, no. 2, pp. 287–294, Feb. 2002.
13. A. Rahman and M. S. Lundstrom, "A compact scattering model for the nanoscale double-gate MOSFET," *IEEE Trans. Electron Devices*, vol. 49, no. 3, pp. 481–489, March 2002.
14. F. Djeffal, Z. Ghoggali, Z. Dibi, and N. Lakhdar, "Analytical analysis of nanoscale multiple gate MOSFETs including effects of hot carrier induced interface charges," *Microelectronics Reliability*, vol. 49, no. 4, pp. 377–381, April 2009.
15. P. Dollfus, and Retailleau, "Thermal noise in nanometric DG MOSFET," *J. of Computational Electronics*, vol. 5, no. 4, pp. 479–482, Dec. 2006.
16. P. Dollfus, "Sensitivity of single and double-gate MOS architectures to residual discrete dopant distribution in the channel," *J. of Computational Electronics*, vol. 5, no. 2–3, pp. 119–123, Sept. 2006.
17. S. Sharma and P. Kumar, "Non overlapped single and double gate SOI/GOI MOSFET for enhanced short channel immunity," *J. of Semiconductor Technology and Science*, vol. 9, no. 3, pp. 136–147, Sept. 2009.
18. A. Kranti, Y. Hao and G. A. Armstrong, "Performance projections and design optimization of planar double-gate SOI MOSFETs for logic technology applications," *Semiconductor Science and Technology*, vol. 23, no. 4, pp. 1–13, 2008.
19. T. C. Lim and G. A. Armstrong, "Scaling issues for analogue circuits using double gate SOI transistors," *Solid State Electronics*, vol. 51, no. 2, pp. 320–327, Feb. 2007.
20. Thomas H. Lee, *The Design of CMOS Radio-Frequency Integrated Circuits*, 2nd Edition, Cambridge University Press, USA, 2004.
21. D. Rechem, S. Latreche, and C. Gontrand, "Channel length scaling and the impact of metal-gate work function on the performance of double-gate MOSFETs," *J. of Physics*, vol. 72, no. 3, pp. 587–599, March 2009.
22. S. Amakawa, K. Nakazato, and H. Mizuta, "A surface potential based cylindrical surrounding gate MOSFET model," *Proc. of Int. Conf. on Solid State Devices and Materials*, Tokyo, Japan, 16–18 Sept. 2003, pp. 1–2.
23. Cong Li, Yiqi Zhuang, and Ru Han, "Cylindrical surrounding-gate MOSFETs with electrically induced source/drain extension," *Microelectronics Journal*, vol. 42, no. 2, pp. 341–346, Feb. 2011.
24. Te Kuang Chiang, "Concise analytical threshold voltage model for cylindrical fully depleted surrounding-gate MOSFET," *Jpn. J. Appl. Phys.*, vol. 44, no. 5, pp. 2948–2952, 2005.
25. H. Kaur, S. Kabra, S. Bindra, S. Haldar, and R. Gupta, "Impact of graded channel design in fully depleted cylindrical/surrounding gate MOSFET for improved short channel immunity and hot carrier reliability," *Solid State Electronics*, vol. 51, no. 3, pp. 398–404, March 2007.
26. Mei Chao Yeh, Zuo Min Tsai, and Huei Wang, "A miniature DC to 50 GHz CMOS SPDT distributed switch," *Proc. of European Symp. on Gallium Arsenide and Other Semiconductor Application*, Paris, 3–4 Oct. 2005, pp. 665–668.
27. Huaxin Lu, Wei Yuan Lu, and Yuan Taur, "Effect of body doping on double-gate MOSFET characteristics," *Semiconductor Science and Technology*, vol. 23, no. 1, pp. 1–6, Jan. 2008.
28. Oana Moldovan, Ferney A. Chaves, David Jimenez, Jean P. Raskin, and Benjamin Iniguez, "Accurate prediction of the volume inversion impact on undoped double-gate MOSFET capacitances," *Int. J. of Numerical Modeling: Electronic Networks, Devices and Fields*, vol. 23, no. 6, pp. 447–457, Nov. 2010.
29. C. Y. Lin, M. W. Ma, A. Chin, Y. C. Yeo, and D. L. Kwong, "Fully silicided N$_i$S$_i$ gate on La$_2$O$_3$ MOSFETs," *IEEE Electron Device Letter*, vol. 24, no. 5, pp. 348–350, May 2003.

30. J. Liu, H. C. Wen, J. P. Lu, and D. L. Kwong, "Dual work-function metal gates by full silicidation of poly-Si with Co-Ni bi-layers," *IEEE Electron Device Letter*, vol. 26, no. 4, pp. 228–230, April 2005.
31. I. V. Singh and M. S. Alam, "Single-gate and double-gate SOI MOSFET structures and compression of electrical performance," *Int. J. of Computer Applications*, vol. 17, pp. 5–11, Oct. 2010.
32. T. Li, C. Hu, W. Ho, H. Wang, and C. Chang, "Continuous and precise work function adjustment for integratable dual metal gate CMOS technology using Hf-Mo binary alloys," *IEEE Trans. on Electron Devices*, vol. 52, no. 6, pp. 1172–1179, June 2005.
33. C. H. Lu, G. Wong, M. Deal, W. Tsai, P. Majhi, J. Chambers, B. Clemens, and Y. Nishi, "Characteristics and mechanism of tunable work function gate electrodes using a bilayer metal structure on SiO_2 and HfO_2," *IEEE Electron Device Letter*, vol. 26, no. 7, pp. 445–447, July 2005.
34. Yuan Taur, D. J. Frank, R. H. Dennard, E. Nowak, P. M. Solomon, and Hon Sum Wong, "Device scaling limits of Si MOSFETs and their application dependencies," *Proc. of IEEE*, vol. 89, no. 3, pp. 259–288, March 2001.
35. S. Kolberg, H. Borli, and T. A. Fjeldly, "Capacitance modeling of short channel double-gate MOSFETs," *Solid State Electronics*, vol. 52, no. 10, pp. 1486–1490, Oct. 2008.
36. Viranjay M. Srivastava, K. S. Yadav, and G. Singh, "Analysis of double-gate CMOS for DP4T RF switch design at 45 nm technology," *J. of Computational Electronics*, vol. 10, no. 1–2, pp. 229–240, June 2011.
37. M. Cheralathan, Antonio Cerdeira, and Benjamin Iniguez, "Compact model for long-channel cylindrical surrounding-gate MOSFETs valid from low to high doping concentrations," *Solid State Electronics*, vol. 55, no. 1, pp. 13–18, Jan. 2011.
38. D. B. Kao, J. P. McVittie, W. D. Nix, and K. C. Saraswat, "Two dimensional thermal oxidation of silicon-I: experiments," *IEEE Trans. on Electron Devices*, vol. 34, no. 5, pp. 1008–1017, May 1987.
39. D. B. Kao, J. P. McVittie, W. D. Nix, and K. C. Saraswat, "Two dimensional thermal oxidation of silicon-II: Modeling stress effects in wet oxides," *IEEE Trans. on Electron Devices*, vol. 35, no. 1, pp. 25–37, Jan. 1988.
40. Vincent Pott, Kirsten E. Moselund, Didier Bouvet, and Adrian Mihai Ionescu, "Fabrication and characterization of gate-all-around silicon nanowires on bulk silicon," *IEEE Trans. on Nanotechnology*, vol. 7, no. 6, pp. 733–744, Nov. 2008.
41. Y. Y. Yeoh, K. H. Yeo, D. W. Kim, S. H. Lee, and C. H. Park, "Characterization of gate-all-around Si-NWFET, including R_{sd}, cylindrical coordinate based 1/f noise and hot carrier effects," *Proc. of IEEE Int. Reliability Physics Symposium*, California, USA, 2–6 May 2010, pp. 94–98.
42. Yun Seop Yu, Namki Cho, Sung Woo Hwang, and Doyeol Ahn, "Analytical threshold voltage model including effective conducting path effect for surrounding-gate MOSFETs with localized charges," *IEEE Trans. on Electron Devices*, vol. 57, no. 11, pp. 3176–3180, Nov. 2010.
43. S. Saurabh and M. Kumar, "Impact of strain on drain current and threshold voltage of nanoscale double-gate tunnel field effect transistor: theoretical investigation and analysis," *Jpn. J. Appl. Phys.*, vol. 48, pp 1–35, June 2009.
44. S. Kolberg, H. Borli, and T. A. Fjeldly, "Modeling, verification and comparison of short channel double-gate and gate-all-around MOSFETs," *Mathematics and Computers in Simulation*, vol. 79, no. 4, pp. 1107–1115, Dec. 2008.
45. F. Djeffal, M. Meguellati, and A. Benhaya, "A two dimensional analytical analysis of subthreshold behavior to study the scaling capability of nanoscale graded channel gate stack DG MOSFETs," *Physica E: Low dimensional Systems and Nanostructures*, vol. 41, no. 10, pp. 1872–1877, Oct. 2009.
46. International Technology Roadmap for Semiconductors-2012, www.public.itrs.net

References

47. M. Reyboz, P. Martin, T. Poiroux, and O. Rozeau, "Continuous model for independent double-gate MOSFET," *IEEE J. of Solid State Circuits*, vol. 53, no. 5, pp. 504–513, May 2009.
48. Ismail Saad and Razali Ismail, "Self aligned vertical double-gate MOSFET with the oblique rotating ion implantation method," *Microelectronics Journal*, vol. 39, no. 12, pp. 1538–1541, Dec. 2008.
49. P. Dutta, B. Syamal, N. Mohankumar, and C. K. Sarkar, "A surface potential based drain current model for asymmetric double-gate MOSFETs," *Solid State Electronics*, vol. 56, no. 1, pp. 148–154, Feb. 2011.
50. F. Djeffal, M. A. Abdi, D. Arar, and T. Bendib, "An analytical subthreshold swing model to study the scalability limits of double-gate MOSFETs including bulk traps effects," *Proc. of 5th Int. Conf. on Design and Technology of Integrated Systems in Nanoscale Era*, Tunisia, 23–25 March 2010, pp. 1–6.
51. Sungmo Kang and Yusuf Leblebichi, *CMOS Digital Integrated Circuits Analysis and Design*, 3rd Edition, McGraw-Hill, New York, USA, 2002.
52. Viranjay M. Srivastava, K. S. Yadav, and G. Singh, "Application of VEE Pro software for measurement of MOS device parameter using C-V curve," *Int. J. of Computer Applications*, vol. 1, no. 7, pp. 43–46, March 2010.
53. Viranjay M. Srivastava, *C-V Measurement Using VEE Pro Software after Fabrication of MOS Capacitance*, 1st Edition, VDM Publishing House, Mauritius, 2010.
54. F. J. Huang and O. Kenneth, "A 0.5 μm CMOS T/R switch for 900 MHz wireless applications," *IEEE J. of Solid State Circuits*, vol. 36, no. 3, pp. 486–492, March 2001.
55. Cong Li, Yiqi Zhuang, and Ru Han, "Cylindrical surrounding-gate MOSFETs with electrically induced source/drain extension," *Microelectronics Journal*, vol. 42, no. 2, pp. 341–346, Feb. 2011.
56. M. Cheralathan and B. Iniguez, "Compact model for long-channel cylindrical surrounding-gate MOSFETs valid from low to high doping concentrations," *Solid State Electronics*, vol. 55, no. 1, pp. 13–18, Jan. 2011.
57. L. Gaionia, M. Manghisonib, L. Rattia, V. Reb, V. Spezialia, and G. Traversib, "Instrumentation for Gate Current Noise Measurements on sub-100 nm MOS Transistors," *Proc. of Topical Workshop on Electronics for Particle Physics*, Greece, Athence, 15–19 Sep 2008, pp. 436–440.
58. Viranjay M. Srivastava, K. S. Yadav, and G. Singh, "Explicit model of cylindrical surrounding double-gate MOSFETs," *WSEAS Trans. on Circuits and Systems*, vol. 12, no. 3, pp. 81–90, March 2013.
59. J. Lee, G. Bosman, K. R. Green, D. Ladwing, "Noise model of gate-leakage current in ultrathin oxide MOSFETs," *IEEE Trans. on Electron Devices*, vol. 50, no. 12, pp. 2499–2506, Dec. 2003.
60. A. J. Scholten, L. F. Tiemeijer, R. J. Havens, and V. C. Venezia, "Noise modeling for RF CMOS circuit simulation," *IEEE Trans. on Electron Devices*, vol. 50, no. 3, pp. 618–632, March 2003.
61. Khaled Ben Ali, Cesar Roda Neve, Ali Gharsallah, and J. P. Raskin, "Impact of crosstalk into high resistivity silicon substrate on the RF performance of SOI MOSFET," *J. of Telecommunications and Information Technology*, vol. 3, no. 4, pp. 93–100, 2010.
62. D. Lederer and J. P. Raskin, "RF performance of a commercial SOI technology transferred onto a passivated HR silicon substrate," *IEEE Trans. on Electron Devices*, vol. 55, no. 7, pp. 1664–1671, July 2008.
63. Serge Gidon, "Double-gate MOSFET modeling," *Proc. of the COMSOL Multiphysics User's Conf.*, Paris, 2005, pp. 1–4.
64. Riza Tamer and Kausik Roy, "Analysis of options in double-gate MOS technology: A circuit perspective," *IEEE Trans. on Electron Devices*, vol. 54, no. 12, pp. 3361–3368, Dec. 2007.
65. Yiming Li and Hong Mu Chou, "A comparative study of electrical characteristic on sub 10-nm double-gate MOSFETs," *IEEE Trans. on Nanotechnology*, vol. 4, no. 5, pp. 645–647, 2005.

66. Pedram Razavi and Ali Orouji, "Nanoscale triple material double gate MOSFET for improving short channel effects," *Proc. of Int. Conf. on Advances in Electronics and Microelectronics*, Valencia, Spain, 29 Sept.-4 Oct. 2008, pp. 11–14.
67. Vaidyanathan Subramanian, "Multiple gate field effect transistor for future CMOS technologies," *IETE Technical Review*, vol. 27, no. 6, pp. 446–454, Dec. 2010.
68. Tae Hun Kim, Hoon Jeong, Ki Whan Song, and Han Park, "A new capacitor-less 1 T DRAM cell: Surrounding gate MOSFET with vertical channel," *IEEE Trans. on Nanotechnology*, vol. 6, no. 3, pp. 352–357, May 2007.
69. S. Venugopalan, Darsen Lu, Yukiya Kawakami, Peter Lee, Ali Niknejad, and Chenming Hu, "BSIM-CG: A compact model of cylindrical/surround gate MOSFET for circuit simulations," *Solid State Electronics*, vol. 67, no. 1, pp. 79–89, Jan. 2012.
70. Usha Gogineni, Jesus Alamo, and Christopher Putnam, "RF power potential of 45-nm CMOS technology," *Proc. of 10th Topical Meeting on Silicon Monolithic Integrated Circuits in RF Systems*, Phoenix, Arizona, USA, 11–13 Jan. 2010, pp. 204–207.

Chapter 6
Hafnium Dioxide-Based Double-Pole Four-Throw Double-Gate RF CMOS Switch

6.1 Introduction

Established radio-frequency complementary metal-oxide-semiconductor (RF CMOS) switch contains MOSFET in its main architecture with 5.0 V of control voltage and requires high value of resistance in circuitry of the transceivers to detect the signal. To avoid the high value of control voltage and resistances, we have designed a novel double-pole four-throw (DP4T) RF switch by using the MOSFET technology and analyzed its performance in terms of drain currents and switching speed in the previous chapters. The reduction in sizing ratio of the gate dielectric, which works as a capacitor in the MOSFET, results in the increase of capacitance and speed of the device. However, this process has reached up to the limit where further reduction of SiO_2 thickness increases the leakage current above the acceptable limit. This problem can be resolved by replacing SiO_2 with materials having high dielectric constants. Hafnium dioxide (also known as Hafnia, HfO_2) is one of them, which has relatively large energy bandgap and a better thermal stability as compared to silicon [1]. It is a leading contender for new high-k gate dielectric films.

When ultrathin gate dielectrics are used in conjunction with poly-silicon gate electrodes, the dopant penetration from the gate into the channel creates a significant problem. Further, difficulty arises with the continued scaling of the physical dimensions of the gate electrode. According to the International Technology Roadmap for Semiconductors (ITRS) [2], the aspect ratio (W/L) for the gate electrode is expected to remain constant as the gate length is scaled down. Generally, the gate sheet resistance has to be maintained at 5 Ω per square area, which implies the need of increase in gate active dopant concentrations as the gate length is scaled down.

The gate materials should be thermodynamically stable on the gate dielectric and must be able to withstand high temperature used in the fabrication of device. In the device fabrication process, high temperatures are used for the activation of dopant atoms in the source, drain, and gate regions of the transistor. Hence, the new gate

Fig. 6.1 Dielectric constant vs. bandgap for gate oxides [23]

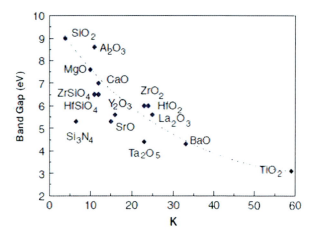

Table 6.1 Properties of hafnium dioxide [9]

S. No.	Properties	Parameters
1	Molecular formula	HfO_2
2	Molar mass	210.49 g/mol
3	Appearance	Off-white powder
4	Density	9.68 g/cm^3, solid
5	Melting point	2,812 °C
6	Boiling point	5,400 °C
7	Solubility in water	Insoluble

electrode material must be chosen with high-k dielectric material, which has been discussed in this chapter.

However, an important consideration in the selection of metal gate electrodes is the work function [3, 4]. In general, the work function of a metal at a dielectric interface is different from its value in the vacuum. This variation needs to be taken into account to design transistor gate stacks. The motivation to replace traditional SiO_2 gate dielectrics with HfO_2 is because it allows increased gate capacitance without affecting the leakage effects. In order to improve the performance of MOSFET devices, HfO_2-based gate layers are being integrated with the achievement of low leakage current (Fig. 6.1). We can also achieve the process of increasing the Debye length and mobility for the switch by the use of HfO_2. Since the HfO_2 has melting point of 2,812 °C [5] (Table 6.1), the designed MOSFET device can work efficiently for high-power switches, jacket water temperature, process temperature, and also for the broadband and carbon nanotube-based nonvolatile random access memory (RAM).

Hafnium dioxide is inorganic, colorless, solid and stable compound of hafnium and also an intermediate which provides Hf metal. It has relatively large energy bandgap. It is an electrical insulator with a bandgap of 5.8 eV [6–9]. HfO_2 is inert and responds with strong acids and strong bases, dissolves slowly in HF acid, and gives fluoro-hafnate anions. The performance of HfO_2 for the MOSFET, such as

6.1 Introduction

oxide capacitance per unit area, threshold voltage, mobility of carriers, drain current, body biasing effect, resistance, capacitance, figure of merit, CMOS switching characteristics with the rise time, fall time, maximum signal frequency, propagation delay, and power dissipation, has been discussed in the following sections.

In this chapter, we have explored the circuit techniques for a DP4T RF CMOS switch, consist of symmetrical independent double-gate (DG) MOSFET having a high dielectric material hafnium dioxide (HfO_2) in place of silicon dioxide (SiO_2). However, the independent gate control in the double-gate devices enhances the circuit performance and robustness while substantially reducing the leakage and chip area at 45-nm technology. We have analyzed various parameters such as drain current, output voltage, effective ON-state resistance, switch ON/OFF ratio, flat-band capacitance, average dynamic power, doping densities, Debye length, mobility of carriers, barrier heights, and insertion loss for the DP4T double-gate (DG) RF CMOS switch. The analysis of this DP4T DG RF CMOS switch with HfO_2 includes the basics of the circuit elements required as integrated circuits for the radio-frequency communication systems. This system provides a plurality of switches, where the power and area could be reduced as compared to the already existing transceiver switch configuration.

However, the enhancement in the frequency response of Si-CMOS devices has motivated their use in the RF/millimeter wave applications, such as high capacity wireless in local area network (WLAN), short range high data rate, wireless personal area network (WPAN), and collision avoidance radar for automobiles [5]. By using Si-CMOS for these applications, we can achieve higher levels of integration and lower cost with improved efficiency [10, 11]. The 65-nm technology has application in 60 GHz power amplifier designs, recently reported research [12–14] has demonstrated 60 GHz power amplifiers in 45-nm technology. It is a leading contender for new high-k dielectric constant. The dielectric performance requires not only precise film thickness control but also very sensitive to compositional or stoichiometric changes (a branch of chemistry that deals with the relative quantities of reactants and products in chemical reactions). This can be achieved with a gate material of hafnium dioxide or hafnium silicate. Hafnium dioxide is a high dielectric, low absorption material usable for coatings in UV (~250 nm) to IR (~10 μm) regions [9].

The bulk MOSFETs show the severe short channel effects (SCE) like drain-induced barrier lowering (DIBL) and threshold voltage roll-off, as the channel length of the device goes down in the nanometer range. DG MOSFETs are good candidates to replace the conventional MOSFETs in this particular region because of their excellent immunity to short channel effects [15–17]. It can be employed with tied gate or independently controlled gate configurations. The back-gate bias can control the threshold voltage in fully depleted silicon-on-insulator (FD SOI) devices with thin buried oxide or in the DG MOSFET [18]. The independent controlling of the front gate and back gate provides wide design and application opportunities such as RF switches and multiplexers. The ability to enhance the gate control and provide logic versatility with a tight physical image and satisfactory leakage characteristics in the double-gate technologies gives ample reasons to pursue research and development activities in this area.

Here, we have used HfO_2 instead of SiO_2 to design a DG MOSFET. A comprehensive study of the RF switch performance for low-power, high-speed DP4T DG RF CMOS switch and structures with HfO_2 to understand the effect of device geometry, and switching properties has been presented. We have investigated the electrical properties of the DG MOSFET, which is very promising for the device miniaturization below 0.1 µm [19]. We have optimized the drain current, discussed the characteristics voltage, and have presented the comprehensive study of the effect of gates in the DP4T RF CMOS switch performance at 45-nm technology in terms of the output voltage. The device structures with different gates are studied to understand the effect of device geometry on RF CMOS with single-gate MOSFET and double-gate MOSFET.

6.2 MOSFET Model with HfO_2

The oxidized HfO_2 films are absorption free for the range between 0.30 and 10 µm [5]. For this HfO_2 films, evaporation causes some dissociation and oxygen loss, which can be recovered with a partial pressure of oxygen during reactive deposition [20]. However, under the low energy evaporation conditions, such as low substrate temperature or with excessive background pressure, the films grow with a porous crystalline microstructure of low packing density and can exhibit index changes when vented to moist air. It is recommended that high-energy deposition techniques such as high substrate temperature are used to decrease the open void volume by increasing the packing density of the microstructure [21, 22].

As compared with planar MOS devices, the double-gate devices exhibit smaller subthreshold and gate leakage currents while offering stronger current drive. The independent biasing of the front gate and back gate in the double-gate technologies has been reported to enhance performance and reduce the chip area due to the reduction of transistor count to implement a given logic function [23]. However, the separate gate access allows for the simplification of circuit topologies and area compactness, both lead to the power and speed improvement in addition to the design flexibility [24–26].

In the symmetrical double-gate devices, very high body doping density with poly-silicon gate or undoped body with near mid-gap metal gate material is used to set the desired threshold voltage (V_{th}). For an asymmetrical n-type DG MOSFET, the front gate and back gate consist of n^+ poly-silicon and p^+ poly-silicon, respectively. For the asymmetrical p-type DG MOSFET, the opposite type of gate is applied, that is, p^+ poly-silicon for the front gate and n^+ poly-silicon for the back gate. The predominant front channel has a significantly lower threshold voltage (V_{th}) and much larger drive current compared with the weak back channel [27–30], which is also discussed in the next section [31]. The front channel can be modulated by back-gate biasing through the gate-to-gate coupling. In the bulk and PD SOI

6.3 Fabrication Process of HfO$_2$-Based DG MOSFET

Fig. 6.2 Schematic of the basic n-type MOSFET (**a**) with HfO$_2$ and (**b**) HfO$_2$ film on Si-substrate

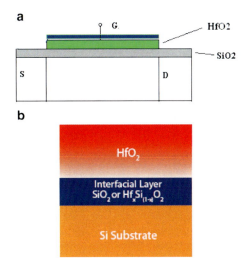

devices, the effectiveness and operating frequency of the well/body bias are limited by the distributed resistance and capacitance (R and C) of the well and body contact. It also tends to degrade with the technology scaling due to the lower body factor in the scaled devices.

Its adhesion is superb to metals such as aluminum and silver. It is an alternative for ultrathin gate oxide in MOSFET due to its high dielectric constant and thermal stability. HfO$_2$ has a melting point of 2,812 °C so it can be used as a refractory material in the insulation of such devices as thermocouples. In the following section, the performance of HfO$_2$ for switch as effective R_{ON}, attenuation, flatband capacitance, average dynamic power, and working efficiency at high temperature are discussed.

The proposed MOSFET with the HfO$_2$ is shown in Fig. 6.2a and the detailed structure of the contact layers used in this design is presented in Fig. 6.2b.

6.3 Fabrication Process of HfO$_2$-Based DG MOSFET

In the deposition of HfO$_2$ films on Si, the HfO$_2$ films have an interfacial layer of either SiO$_2$ or Hf$_x$Si$_{(1-x)}$O$_2$ that can substantially change the overall dielectric property of the film. The crystallization of HfO$_2$ occurs at 400–450 °C causing grain boundary leakage current and nonuniformity of the film thickness. As a result, impurities such as Oxygen, Boron, and Phosphorous can penetrate the grain boundaries during the high temperature postprocessing. It causes equivalent oxide thickness scaling and reliability concerns when Hf-based high-k ultrathin gate oxides are integrated into the high-temperature CMOS processes [23]. Nitrogen introduction into the HfO$_2$ films has significantly improved the electrical as well as crystalline properties.

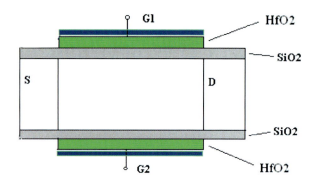

Fig. 6.3 Schematic of n-type DG MOSFET with HfO$_2$

Table 6.2 Dielectric constant, bandgap, and conduction band offset on Si of the candidate gate dielectrics

Material	K	Bandgap (eV)	Offset (eV)
Si	11.7	1.1	–
SiO$_2$	3.9	9	3.2
Al$_2$O$_3$	9	8.8	2.8
HfO$_2$	25	5.8	1.4

We have already designed the SiO$_2$-based DG MOSFET as shown in Fig. 3.1 in Chap. 3. The HfO$_2$ adhesion property is superb to metals such as aluminum and silver. So, the aluminum is used in this discussion as shown in Fig. 6.3 with light blue color and HfO$_2$ region is shown with green color. Here we can sputtered the Hf metal film on SiO$_2$ upon a substrate wafer and then dipped it into the HNO$_3$ to form high-k HfO$_2$. It is used for semiconductor devices as a gate dielectric and forms the capacitor dielectric, so that the equivalent parasitic capacitance will decrease for DG MOSFET which is a series combination, in view of gate-1 (G$_1$) and gate-2 (G$_2$). Now, aluminum layer is metalized onto the surface of HfO$_2$ for the gate pattern. A change in the temperature directs to distinction in properties of the bulk-Si and the Si–SiO$_2$ interface which causes the variations in saturation current [32]. This saturation current is a combination of electron which is generated from interface states, tunneled through the thinner oxide, and the fractional electron is trapped and de-trapped in the HfO$_2$ layer.

As the temperature increases, the voltage drop across the insulator increases, whereas across the depletion region it decreases [33]. However, the voltage drop across the SiO$_2$ layer is larger than that across the HfO$_2$ layer. So the current mechanism in HfO$_2$ layer becomes important when temperature increases. However, HfO$_2$ exhibits reduced channel nobilities and larger leakage currents relative to pure SiO$_2$ [34]. The bandgap, E_g, increased from 5.52 eV for Hf silicate with an Hf/Si ratio of 3:1 to 6.61 eV for Hf silicate with an Hf/Si ratio of 1:3. The characteristics for the SiO$_2$ and HfO$_2$ are given in Table 6.2.

6.4 Parameters of HfO$_2$-Based MOSFET

There are various parameters for hafnium dioxide-based MOSFET. We have discussed some of major parameters available in the device related to the transceiver switches.

6.4.1 Oxide Capacitance per Unit Area

It is a capacitance of oxide layer with high-k dielectric material. The value of C_{ox} determines the amount of electrical coupling that exists between the gate electrode and the p-type silicon region. It is defined as follows:

$$C_{OX} = \frac{\varepsilon_{OX}}{t_{OX}} \tag{6.1}$$

For SiO$_2$, $\varepsilon_{ox} = 3.9\varepsilon_o$ and for HfO$_2$, $\varepsilon_{ox} = 25\varepsilon_o$. So, using (6.1), we found that the C_{ox} value is 0.35 µF/cm^2 for SiO$_2$ and 2.21 µF/cm^2 for HfO$_2$. However, sufficiently high value for hafnium dioxide makes it better in order to maintain the high value of current due to accumulation of more charges in the channel.

6.4.2 Threshold Voltage

The value of gate-to-source voltage needed to cause surface inversion is called the threshold voltage. At room temperature, for a given doping concentration, the threshold voltage is given by:

$$V_{th} = \frac{1}{C_{ox}} \sqrt{2q\varepsilon N(2\phi)} \tag{6.2}$$

This threshold voltage is 0.63 V for n-type MOSFET and −1.54 V for p-type MOSFET in case of silicon as a substrate and SiO$_2$ as an oxide layer while for HfO$_2$ these values are 0.58 V and 1.12 V, respectively. However, the lower value of threshold voltage is advantageous for the high-speed switching applications.

6.4.3 Drain Currents

The bandgap of SiO$_2$ is 9 eV, whereas for HfO$_2$ it is 5.8 eV. Hence the mobility of carriers is more in HfO$_2$ as compared to SiO$_2$. The current–voltage equations of the n-channel MOSFET is given as follows:

$$I_d = 0, \quad V_{gs} < V_{th} \tag{6.3.a}$$

$$I_{d(\text{lin})} = \frac{\mu_n C_{ox}}{2} \frac{W}{L} \left[2 \cdot (V_{gs} - V_{th}) V_{ds} - V_{ds}^2 \right], \quad V_{ds} < V_{gs} - V_{th} \quad (6.3.\text{b})$$

$$I_{d(\text{sat})} = \frac{\mu_n C_{ox}}{2} \frac{W}{L} (V_{gs} - V_{th})^2, \quad V_{ds} \geq V_{gs} - V_{th} \quad (6.3.\text{c})$$

However, the drain current in all conditions is directly proportional to the mobility and oxide capacitance per unit area and as these values are larger for HfO_2 as compared to SiO_2, large amount of current is achieved by using HfO_2 for the same applied voltages and aspect ratio.

6.4.4 Body Bias Effect

This effect occurs when the voltage V_{sb} exists between the source and bulk terminals of a MOSFET. The body bias voltage increases to the threshold voltage of the device. Hence, its low value is preferred. The body bias effect coefficient is given by:

$$\gamma = \frac{\sqrt{2q\varepsilon N \phi}}{C_{ox}} \quad (6.4)$$

Its value is 0.053 for silicon and 0.02 for hafnium. Thus hafnium is preferred for low bias effect.

6.4.5 Resistances

The drain-source resistance for a MOS is inversely proportional to the carrier mobility and the oxide capacitance per unit area by the relation:

$$R_{ds} = \frac{1}{\mu C_{ox} \left(\frac{W}{L}\right)(V_{dd} - V_{th})} \quad (6.5)$$

Due to the higher value of C_{ox} and μ, HfO_2 offers lesser resistance to the current flow as compared to SiO_2.

6.4.6 Capacitances

MOSFET capacitances are given by the relation:

$$C_g = C_{ox} WL \quad (6.6.\text{a})$$

6.5 Switching Characteristics of HfO$_2$-Based MOSFET

$$C_{gs} = C_{gd} = \frac{C_g}{2} \qquad (6.6.b)$$

For a chosen dimension of the device, C_g is around six times higher for HfO$_2$ than SiO$_2$; thus, it is providing better insulation to the gate from the channel resulting in higher input impedance and hence suitable for high-gain amplifier applications.

6.4.7 Figure of Merit

An indication of the frequency response may be obtained from the parameter ω_0, as:

$$\omega_0 = \frac{g_m}{C_g} = \frac{\mu}{L^2}(V_{gs} - V_{th}) \qquad (6.7)$$

where g_m is transconductance and C_g is the gate capacitance.

This reveals that the switching speed depends on gate voltage above the threshold voltage as well as on the carrier mobility and inversely proportional to the square of channel length. Thus, a fast circuit requires g_m being as high as possible. The higher value is achieved with the use of hafnium due to reduced value of threshold voltage and higher mobility of carriers as compared to silicon.

6.5 Switching Characteristics of HfO$_2$-Based MOSFET

In the previous sections we have designed the hafnium dioxide-based MOSFET and their characteristics related to the switching operation are discussed as follows:

6.5.1 Fall Time

It is a time for a signal to fall from 90 % to 10 % of its magnitude; however, sometimes, it may fall from 100 % to 0 %. For a given output capacitance, the fall time is calculated by:

$$T_f = 2.2\, R_n C_{load} \qquad (6.8.a)$$

where R_n and C_{load} are the circuitry resistance and load capacitance, respectively, for the fall time. As the value of the resistance is lesser for hafnium, low value of fall time is attained. For an ideal switch, fall time should be zero.

6.5.2 Rise Time

It is a time for a signal to rise from 10 % to 90 % of its magnitude; however, sometimes it may rise from 0 % to 100 %. For a given output capacitance, the rise time is calculated by:

$$T_r = 2.2 \, R_p C_{load} \tag{6.8.b}$$

where R_p and C_{load} are the circuitry resistance and load capacitance, respectively, for the rise time. As the value of the resistance is lesser for hafnium, low value of rise time is attained. Thus faster switching is achieved with Hf as substrate and HfO_2 as the oxide layer. For an ideal switch, rise time should be zero.

6.5.3 Maximum Signal Frequency

This is the largest frequency that can be applied to the gate and still allows the output to settle to a definable state. This is calculated by:

$$F_{max} = \frac{1}{(T_r + T_f)} \tag{6.8.c}$$

With lower values of T_r and T_f the value of F_{max} increases to a greater extent and hence the operating range using hafnium.

6.5.4 Propagation Delay

The simplest approach for calculating the propagation delay times is based on the estimation of the average capacitive current during charging and discharging. The digital devices with smaller delay times are better because it can operate at higher frequencies which are essential in modern digital circuits. The propagation delay is calculated from the expression:

$$T_p = 0.35(R_n + R_p) \, C_{load} \tag{6.8.d}$$

For a given output capacitance, the lower delay is achieved using hafnium due to lower value of time constants.

6.6 DP4T Switch Design with HfO$_2$-Based DG MOSFET

Table 6.3 Comparison of parameters of HfO$_2$-based MOSFET with the SiO$_2$-based MOSFET

S. No.	Parameters	SiO$_2$-based MOSFET	HfO$_2$-based MOSFET
1	Oxide capacitance	0.35 µF/cm^2	2.21 µF/cm^2
2	n-MOSFET threshold voltage	0.63 V	0.58 V
3	p-MOSFET threshold voltage	−1.54 V	−1.12 V
4	Drain to source current	Less	More
5	Mobility	Less	More
6	Body biasing effects	0.053	0.020
7	Resistance between drain and source	Low	High
8	Capacitance between drain and source	C	6C
9	Figure of merit	Low	High
10	Rise time	High	Low
11	Fall time	High	Low
12	Maximum signal frequency	Low	High
13	Propagation delay	High	Low
14	Power dissipation	Low	High

6.5.5 Power Dissipation

The total power dissipation in the MOSFET circuit is due to both static power and dynamic power. As the high value of drain current is achieved in hafnium, the static power dissipation is high and it is given by:

$$P_{dc} = V_{dd} I_{ddQ} \qquad (6.9.a)$$

For a given supply voltage and output capacitance the dynamic power dissipation depends upon the frequency of operation by the relation:

$$P_{dyn} = V_{dd}^2 C_{load} f \qquad (6.9.b)$$

This reveals that the fast circuit dissipates more power.

Above discussed parameters are summarized in Table 6.3. The numerical values of the above parameters depend on the values taken into account for the observations.

6.6 DP4T Switch Design with HfO$_2$-Based DG MOSFET

The modulated signal is transmitted through a switch and the switch makes its way to the antenna for releasing into space. This modulated signal is received by the antenna and makes its way through the switching path to the receiver. A single-pole double-throw (SPDT) is the fundamental switch that links between one antenna and the transmitter/receiver. Due to the single operating frequency, this type of SPDT

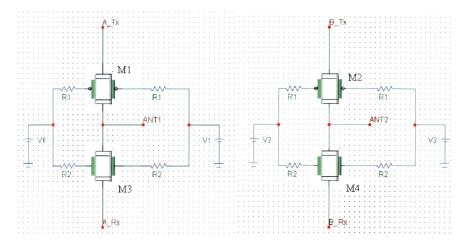

Fig. 6.4 DP4T RF CMOS switch with HfO$_2$-layered double-gate MOSFET

switch has a limited data transfer rate. Therefore, a proposed DP4T switch design is useful to enhance the switch performance for MIMO applications. This DP4T switch can send and receive two parallel data streams simultaneously [35, 36].

A symmetric DG MOSFET results when two gates have the same work function and a single input voltage is applied to both gates. An asymmetric DG MOSFET either has synchronized but different input voltages to two identical gates or has the same input voltage to two gates that have distinct work functions [36]. To avoid the application of these two gate and respective voltages, a DP4T RF CMOS switch is a useful device.

A specification for the RF CMOS switch includes the maximum drain saturation current, common-source forward transconductance, operating frequency, and output power. According to the earlier reported literature [37], DP4T switch is a fundamental switch for MIMO applications because parallel data streams can be transmitted or received simultaneously using the multiple antennas. These switches can be used with digital and baseband analog systems, which are controlled by on-chip digital and analog signals in the design [38, 39]. This switch contains CMOS in its architecture and needs only two control lines (V_1, V_2) of 1.2 V to control the signal traffic between two antennas and four ports as shown in Fig. 6.4. Hence, it improves the port isolation performance two times as compared to the double-pole double-throw (DPDT) switch and reduces the signal distortion [25, 40, 41]. Moreover, the signal fading effects can be reduced because sending the identical signals through multiple antennas will most likely result in a high-quality combined signal at the receiver end. For the design of DP4T DG RF CMOS switch with HfO$_2$, we design an n-type DG MOSFET with HfO$_2$ as shown in Fig. 6.3 and then use this to design DP4T RF CMOS switch with HfO$_2$ layered DG MOSFET as shown in Fig. 6.4. Similarly, we can design p-type DG MOSFET. Since four transistors are used for two antennas and by using the CMOS functionality, at a time any one of transistor M_1 (p-MOSFET) or M_3 (n-MOSFET) will operate and in

6.7 Characteristics of DP4T Switch with HfO$_2$-Based DG MOSFET 155

Table 6.4 Working functionality of DP4T RF CMOS switch with HfO$_2$ DG MOSFET

Supply voltages		Transceiver A		Transceiver B	
V_1	V_2	M_1 (p-MOS)	M_3 (n-MOS)	M_2 (p-MOS)	M_4 (n-MOS)
Low	Low	ON	OFF	ON	OFF
Low	High	ON	OFF	OFF	ON
High	Low	OFF	ON	ON	OFF
High	High	OFF	ON	OFF	ON

the same fashion any one of transistor M_2 (p-MOSFET) or M_4 (n-MOSFET) will operate. However, detailed working functionality is given in Table 6.4.

For example, at an instance, the transmitted signal from power amplifier (PA) is sent to the transmitter "A" which is shown in Fig. 6.4 named as "A_Tx" port and travels to the ANT$_1$ while the received signal will travel from the ANT$_2$ to the receiver "B" named as "B_Rx" port and passes onto the low-noise amplifier or any other application. So it has two poles and four terminals for data transmitting and receiving.

6.7 Characteristics of DP4T Switch with HfO$_2$-Based DG MOSFET

The presented compact model accounts for the charge quantization within the channel, Fermi statistics, and non-static effects in the transport model. The main theme of this compact switch model are as follows:

(a) The DIBL is minimized by the shielding effect of the DG MOSFET, which allows reduction in the channel length from 45 nm.
(b) The device transconductance per unit width is maximized by the combination of the DG MOSFET and by a strong velocity overshoot, which occurs in response to the abrupt variation of the electric field at the source end of the channel [42].
(c) Increase in the device transconductance per unit width can be further strengthened near the drain in view of the short device length.

As a result, the sustained electron velocity of nearly twice the saturation velocity is possible. The following observations proved the potential performance advantages of the double-gate device architecture as a switch. The various parameters associated with the proposed design of a MOSFET are calculated for making a comparison between SiO$_2$ and HfO$_2$ as an oxide material [23, 33]. After the simulation of the device, we have analyzed the following parameters for this DP4T switch:

6.7.1 Drain Current Analysis

For the proposed DP4T switch with HfO$_2$-based DG MOSFET, the drain current is described with the idea of Pao and Sah phenomena [43, 44] that includes both the drift and diffusion transport tendencies in the silicon film, resulting in a current description with flat transitions between the linear and saturation operating regions. Under the approximation that the mobility is independent of the position in the channel, the drain current I_d for a MOSFET can be expressed as [43, 44]:

$$I_d = \mu \frac{W}{L} \int_0^{V_{ds}} Q_1 \, dV \tag{6.10}$$

where μ, W, and L are the effective electron mobility, channel width, and effective channel length, respectively. Q_1 is the total (integrated in the transverse direction) inversion charge density inside the silicon film of a symmetric DG MOSFET. At a given location x is defined as:

$$Q_1 = -2q \int_0^{T_{Si}} (n - n_i) \, dx = -2q \int_{\phi_0}^{\phi_s} \frac{n - n_i}{F} \, d\phi \tag{6.11}$$

where F is the electric field. Since there is no fixed charge in the undoped body, Q_1 can be taken as being the total charge in the semiconductor [44]:

$$Q_1 = 2\varepsilon_s F_s = -2C(V_{GF} - \phi_s) \tag{6.12}$$

where F_S is the electric field at the surface, and the factor of 2 comes from the symmetry of double-gate MOSFET. An equivalent to the Pao and Sah's equation [43, 44] for the SOI MOSFET may be obtained by substituting (6.12) into (6.11), which yields the following generalized two integral formulations for drain current [45]:

$$I_d = 2\mu \frac{W}{L} \int_0^{V_{ds}} \int_{\phi_0}^{\phi_s} \frac{qn}{F} \, d\phi \, dV \tag{6.13}$$

with $n = n_i \, e^{\beta(\phi - V)}$. Hence, we can conclude that for the device under test, charge Q_1 in the single-gate (SG) MOSFET is double than the DG MOSFET. So the current will be double in DG MOSFET as compared to the SG MOSFET.

6.7.2 ON/OFF Ratio and Insertion Loss

A single switch element is characterized for ON/OFF ratio and insertion loss. The ON/OFF ratio of a single switch element is: $10 \log(2\pi R_{ON} C_{OFF} f_0)$ where R_{ON},

6.7 Characteristics of DP4T Switch with HfO₂-Based DG MOSFET

Fig. 6.5 ON/OFF ratio for the proposed DP4T RF CMOS switch

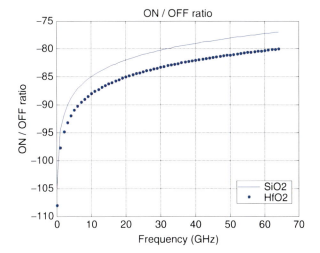

C_{OFF}, and f_0 are ON-resistance, OFF capacitance, and the frequency, respectively. So for the DG MOSFET switch at the high frequency (GHz range) this ratio is higher, means once again switching become fast for the HfO₂-based DP4T DG RF CMOS switch. The insertion loss is given by:

$$\left(\frac{R_{ON} + 2Z_0}{2Z_0}\right) \quad (6.14)$$

where Z_0 is a fixed characteristic impedance and taken as 50 Ω as shown in Fig. 6.5. R_{ON} is the resistance of device at ON-state. For the DG MOSFET, ON-state resistance becomes $R_{ON}/2$ (parallel combination of R_{ON} due to front gate and back gate). So the insertion loss for the DP4T DG RF CMOS switch becomes less as compared to SG MOSFET switch.

6.7.3 ON-Resistance (R_ON) and Attenuation

In the DP4T switch, an effective ON-state resistance R_{ON} is defined as:

$$R_{ON} = \frac{1}{\mu C_{ox} \frac{W}{L} (V_{gs} - V_{th})} \quad (6.15)$$

This R_{ON} should be less for a good RF switch and it is inversely proportional to the intrinsic transconductance. This transconductance is higher in HfO₂ compared to SiO₂, so it is better to use HfO₂ for a DP4T switch. The level of attenuation (ATT) is given by the following expression [46]:

Fig. 6.6 Attenuation for the proposed DP4T RF CMOS switch with respect to the applied control voltage

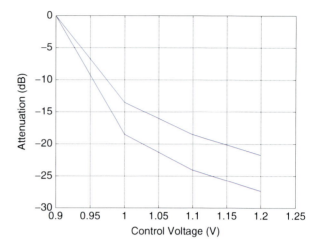

Fig. 6.7 Insertion loss for the proposed DP4T RF CMOS switch with the ON-state resistance

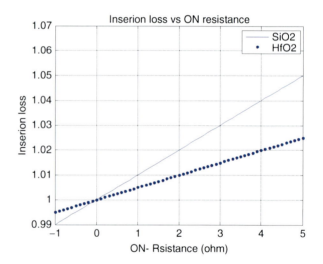

$$\text{ATT} = 20 \cdot \log\left(1 + \frac{R_{\text{ON}}}{2Z_0}\right) \tag{6.16}$$

where R_{ON} is the ON-state series resistance of the attenuator modeled using (6.15) and Z_0 is 50 Ω as shown in Figs. 6.6 and 6.7. This attenuation is directly proportional to R_{ON}, so the attenuation improves by using HfO_2-based DG MOSFET.

6.7.4 Flat-Band Capacitance and Dynamic Power

Since the dielectric constant of HfO$_2$ is 25 and for SiO$_2$ it is 3.9, so for flat-band condition, the flat-band capacitance is given as follows:

$$C_{FB} = \frac{C_{OX} \cdot \varepsilon_{Si}.A/\lambda_D}{C_{OX} + \varepsilon_{Si}.A/\lambda_D} \tag{6.17}$$

where λ_D is the extrinsic Debye length, oxide capacitance, C_{OX}, is 5.202 pF, dielectric permittivity of the substrate material, ε_{si}, is $11.7 \times 8.85 \times 10^{-14}$ Fcm^{-1}, and A is the gate area in cm^2, so we obtained the value of λ_D 1.279×10^{-5} cm as well as C_{FB} 3.17 pF for Si and 4.72 pF for Hf. This high flat-band capacitance is useful for DP4T switch. For the higher flat-band capacitance, at a fixed charge, reduced gate voltage is required as compared to the lower flat-band capacitance. So at low voltage, using the hafnium dioxide, we can transmit or receive data at lower supply voltage.

For CMOS circuits where no DC current flows, average dynamic power, $P_{avg} = C_L \times V_{DD}^2 \times f$, for HfO$_2$ bandgap is 6.0 eV and for SiO$_2$ it is 8.9 eV, so the average dynamic power due to V_{DD} for HfO$_2$ will be less, which is better for a DP4T RF CMOS switch. However, several problems arise when using hafnium dioxide or hafnium silicate materials in the device. One of them is to find p-type and n-type gate metals to match with the valence and conduction band edges of silicon [47]. This problem can be rooted with the band alignment and its controlling mechanisms between the materials in the gate stack.

6.7.5 Debye Length Calculation and Mobility

The Debye length is become significant in the modeling of solid state devices as improvements in lithographic technologies have enabled smaller geometries. The Debye length is given as follows [48]:

$$L_D = \sqrt{\frac{\varepsilon_{Si} k_B T}{q^2 N_d}} \tag{6.18}$$

where ε_{Si}, k_B, N_d, q, and T are the dielectric constant of the material used, Boltzmann's constant, density of donors in a substrate, electronic charge, and absolute temperature, respectively. From Table 6.1, Si has dielectric constant of 3.9 and Hf has 25. So we found that $L_D(Si) < L_D(Hf)$. It means a large distance, over which significant charge separation can occur, is available with Hf compared to Si.

The mobility of carrier for this combination of HfO$_2$ is given by:

$$\mu(T) = \mu_0 e^{-\beta T} \tag{6.19}$$

where μ_0 and β are the carrier mobility and a constant describing the sensitivity of mobility to temperature, respectively. The bandgap of SiO_2 is 9.0 eV, whereas for HfO_2 it is 5.8 eV. So the mobility is more in HfO_2 as compared to SiO_2.

6.7.6 Potential Barrier

The potential barrier exists between the metal and semiconductor layer when they are in close contact, this stops the majority of charge carriers to pass from one layer to the other layer. Only a few charge carriers have an adequate amount of energy to pass through the barrier and cross to the other side of the material. Since the barrier height is the property of a material, we try to use these materials for the CMOS in application of RF, whose barrier height is small. Here is a possibility to create an alloy between metal and semiconductor junction, at the time of annealing, which lowers the barrier height [49, 50]. The probability of tunneling becomes high for extremely thin barriers (in the tens of nanometers). By the heavy doping process one can make the very thin barrier which approximately has a concentration of 10^{19} dopant atoms/cc or more. As the barrier height is closer to zero, the ohmic contact increases. On the basis of Debye length as well as the mobility and potential barrier, we found that HfO_2 is a suitable material for the designing of DG MOSFET as an application of DP4T RF CMOS switch. For Si-substrate, with Boron doping concentration of 10^{19} atoms/cc, mobility is 70.90 cm^2/V^{-sec} and resistivity will be 8.8×10^{-3} $\mu\Omega/cm$.

6.8 Conclusions

The motivation to replace traditional SiO_2 gate dielectrics with HfO_2 is mainly for the requirement of high-k dielectrics and latest development in Hf-based high-k dielectrics which satisfies the process of DP4T RF CMOS switch. In order to improve the integration and performance of CMOS devices, and its applications, Hf-based gate layers are being integrated into DG MOSFET to achieve low leakage current. Excellent gate transistors with improved performance based on Hf-based gate dielectrics as the insulating layers are achieved. After designing the DP4T RF CMOS switch with the structure of DG MOSFET, using HfO_2, we have analyzed the performance of the switch. It includes the basics of the circuit elements parameter required for the radio-frequency subsystems of the integrated circuits such as drain current, threshold voltage, ON/OFF ratio, insertion loss, ON-resistances, attenuation, capacitances, dynamic power, Debye length, mobility of carriers, barrier heights, and switching speed.

Moreover, an analytic threshold voltage model is presented for the SG MOSFET and DG MOSFET. Device simulations are also done and the results obtained are compared with the theoretical analysis for the model. For the purpose of RF switch, we achieve the process to minimize control voltage and resistance for the switch ON condition and also we achieve the process to increase the Debye length and mobility for the switch with the application of HfO_2. Since the HfO_2 has a melting point of 2,812 °C, the designed DP4T switch can work sufficiently for high-power switches, Jacket water temperature, and process temperature and also for broadband and carbon nanotube-based nonvolatile RAM, high-energy radiotherapy applications, analog to digital converters, and GSM applications [51–53]. With the increase of control voltage or gate voltage the drain current increases for the CMOS circuit, designed with hafnium dioxide.

References

1. T. Manku, "Microwave CMOS device physics and design," *IEEE J. of Solid State Circuits*, vol. 34, no. 3, pp. 277–285, March 1999.
2. International Technology Roadmap for Semiconductors (ITRS), www.itrs.net
3. A. Cerdeira, B. Iniguez, and M. Estrada, "Compact model for short channel symmetric doped double-gate MOSFETs," *Solid State Electronics*, vol. 52, no. 7, pp. 1064–1070, July 2008.
4. Behzad Razavi, *RF Microelectronics*, 3rd Edition, Prentice Hall, New Jersey, 1998.
5. H. S. Baik and S. J. Pennycook, "Interface structure and non-stoichiometry in HfO_2 dielectrics," *IEEE Applied Physics Letter*, vol. 85, pp. 672–674, 2009.
6. Viranjay M. Srivastava, K. S. Yadav, and G. Singh, "Analysis of drain current and switching speed for SPDT switch and DPDT switch with proposed DP4T RF CMOS switch," *J. of Circuits, Systems and Computers*, vol. 21, no. 4, pp. 1–18, June 2012.
7. J. C. Lee, "Single-layer thin HfO_2 gate dielectric with n^+ poly-Silicon," *Proc. of IEEE Symposium on VLSI Technology*, Honolulu, Hawaii, USA, 13–15 June 2000, pp. 44–45.
8. Tingting Tan, Zhengtang Liu, Hongcheng Lu, Wenting Liu, Feng Yan, and Wenhua Zhang, "Band structure and valence band offset of HfO_2 thin film on Si substrate from photoemission spectroscopy," *J. of Applied Physics A: Materials Science and Processing*, vol. 97, no. 2, pp. 475–479, 2009.
9. N Shashank, S Basak, and R.K Nahar, "Design and simulation of nano scale high-k based MOSFETs with poly Silicon and metal gate electrodes," *Int. J. of Advancements in Technology*, vol. 1, No. 2, pp. 252–261, 2010.
10. Yuan Taur, "CMOS Scaling into the Nanometer Regime," *Proc. of IEEE Journal*, vol. 85, no. 4, pp. 486–504, April 1997.
11. Riza Tamer and Kausik Roy, "Analysis of options in double-gate MOS technology: A circuit perspective" *IEEE Trans. on Electron Devices*, vol. 54, no. 12, pp. 3361–3368, Dec. 2007.
12. Kim Keunwoo, Ching Te Chuang, J. B. Kuang, and K. J. Nowka, "Low-power high-performance asymmetrical double-gate circuits using back-gate controlled wide tunable range diode voltage," *IEEE Trans. on Electron Devices*, vol. 54, no. 9, pp. 2263–2268, Sept. 2007.
13. A. M. Street, "RF switches design," *The Institution of Electrical Engineers*, London, vol. 4, pp. 1–7, 2000.
14. Viranjay M. Srivastava, K. S. Yadav, and G. Singh, "Design and performance analysis of double-gate MOSFET over single-gate MOSFET for RF switch," *Microelectronics Journal*, vol. 42, no. 3, pp. 527–534, March 2011.
15. Gary K. Yeap, Farid N. Najm, *Low power VLSI design and technology*, Pearson Addison Wesley, 1st Edition, 2008.
16. T. H. Lee, *The Design of CMOS Radio Frequency Integrated Circuits*, Cambridge University Press, 2nd Edition, 2004.

17. Eric Pop, "Energy dissipation and transport in nanoscale devices," *J. of Nano Research*, vol. 3, no. 3, pp. 147–169, 2010.
18. Viranjay M. Srivastava, K. S. Yadav, G. Singh, "Design and performance analysis of cylindrical surrounding double-gate MOSFET," *Microelectronics Journal*, vol. 42, no. 10, pp. 1124–1135, Oct. 2011.
19. Viranjay M. Srivastava, K. S. Yadav, and G. Singh, "DP4T RF CMOS switch: A better option to replace SPDT switch and DPDT switch," *Recent Patents on Electrical and Electronic Engineering*, vol. 5, no. 3, pp. 244–248, Oct. 2012.
20. 'Aluminum oxide, Al_2O_3 for optical coating: A product catalogue' MATERION Advanced Chemicals, USA, 2008.
21. Thomas Sokollik, "Investigations of Field Dynamics in Laser Plasmas with Proton Imaging," *Plasma Physics*, vol. 1, pp. 17–24, 2011.
22. Ta Chang Tien, Li Chuan Lin, Lurng Shehng Lee, Chi Jen Hwang, Siddheswar Maikap, and Yuri M. Shulga, "Analysis of weakly bonded oxygen in $HfO_2/SiO_2/Si$ stacks by using HRBS and ARXPS," *J. of Material Science: Material Electronics*, vol. 21, no. 5, pp. 475–480, 2010.
23. A. P. Huang, Z. C. Yang, and Paul K. Chu, "Hafnium based high-k gate dielectrics," *Advances in Solid State Circuits Technologies*, pp. 333–350, April 2010.
24. Yuhua Cheng, M. Deen, and Chih Chen, "MOSFET modeling for RF IC design," *IEEE Trans. on Electron Devices*, vol. 52, no. 7, pp. 1286–1303, July 2005.
25. S. Panda and M. Ray Kanjilal, "Thermal and Flicker noise modeling of a double gate MOSFET," *J. of Advances in Power Electronics and Instrumentation Engineering*, vol. 148, no. 1, pp. 43–49, 2011.
26. Rajeev Sharma, Sujata Pandey and Shail Bala Jain, "Compact modeling and simulation of nanoscale fully depleted DG-SOI MOSFETS," *J. of Computational Electronics*, vol. 10, no. 1–2, pp. 201–209, 2011.
27. E. J. Nowak, I. Aller, T. Ludwig, K. Kim, R. V. Joshi, C. T. Chuang, K. Bernstein, and R. Puri, "Turning Silicon on its edge," *IEEE Circuits Devices Mag.*, vol. 20, no. 1, Jan./Feb. 2004, pp. 20–31.
28. K. Kim and J. G. Fossum, "Double-gate CMOS: Symmetrical versus asymmetrical-gate devices," *IEEE Trans. on Electron Devices*, vol. 48, no. 2, pp. 294–299, Feb. 2001.
29. Jente B. Kuang, Keunwoo Kim, Hung C. Ngo, Fadi H. Gebara, and Kevin J. Nowka, "Circuit techniques utilizing independent gate control in double-gate technologies," *IEEE Trans. on Very Large Scale Integration Systems*, vol. 16, no. 12, pp. 1657–1665, Dec. 2008.
30. J. W. Han, C. J. Kim, and Y. K. Choi, "Universal potential model in tied and separated double-gate MOSFET with consideration of symmetric and separated asymmetric structure," *IEEE Trans. on Electron Devices*, vol. 55, no. 6, pp. 1472–1479, June 2008.
31. S. M. Sze, *Semiconductor Devices: Physics and Technology*, 2nd Edition, Tata McGraw-Hill, India, 2004.
32. 'A Product catalogue,' READE Advances Materials, England, 2005.
33. C. Wang and J. Hwu, "Characterization of stacked Hafnium-oxide (HfO_2) / Silicon-dioxide (SiO_2) metal-oxide-semiconductor tunneling temperature sensors" *J. of Electrochemical Society*, vol. 157, no. 10, pp. 324–328, 2010.
34. M. Fadel, and O. Azim, "A study of some optical properties of hafnium dioxide (HfO_2) thin films and their applications," *J. of Applied Physics Materials Science and Processing*, vol. 66, no. 3, pp. 335–343, 1997.
35. Kaushik Roy, "Design of high performance sense amplifier using independent gate control in sub-50 nm double gate MOSFET," *Proc. of Int. Symp. on Quality Electronic Design*, San Jose, CA, USA, 21–23 March 2005, pp. 490–495.
36. Oana Moldovan, Ferney A. Chaves, David Jimenez, Jean P. Raskin, and Benjamin Iniguez, "Accurate prediction of the volume inversion impact on undoped double-gate MOSFET capacitances," *Int. J. of Numerical Modeling: Electronic Networks, Devices and Fields*, vol. 23, no. 6, pp. 447–457, Nov. 2010.

References

37. S. Sanayei and N. Aria, "Antenna selection in MIMO systems," *IEEE Communications Magazine*, Oct. 2004, pp. 68–73.
38. Viranjay M. Srivastava, K. S. Yadav, and G. Singh, "Optimization of drain current and voltage characteristics for the DP4T double-gate RF CMOS switch at 45-nm technology," *Procedia Engineering*, vol. 38, pp. 486–492, April 2012.
39. R. H. Caverly, S. Smith, and J. Hu, "RF CMOS cells for wireless applications," *J. of Analog Integrated Circuits and Signal Processing*, vol. 25, no. 1, pp. 5–15, 2001.
40. Viranjay M. Srivastava, K. S. Yadav, and G. Singh, "Analysis of Double-Pole Four-Throw RF CMOS Switch with HfO_2," *Proc. of Nat. Symp. on Microwave Processing of Materials*, India, 28 Nov. 2010, p. 24.
41. D. Yamane, H. Seita, W. Sun, S. Kawasaki, H. Fujita, and H. Toshiyoshi, "A 12-GHz DPDT RF-MEMS switch with layer-wise waveguide/actuator design technique," *Proc. of 22nd IEEE Conf. on Micro Electro Mechanical Systems*, Sorrento, USA, 25–29 Jan 2009, pp. 888–891.
42. Giorgio Baccarani and Susanna Reggiani, "A compact double-gate MOSFET model comprising quantum-mechanical and nonstatic effects," *IEEE Trans. on Electron Devices*, vol. 46, no. 8, pp. 1656–1666, Aug. 1999.
43. Hugues Nurray, patrik martin, and Serge Bardy, "Taylor expansion of surface potential in MOSFET: application to Pao-Sah integral," *Active and Passive Electronic Components*, vol. 2010, pp. 1–11, 2010.
44. Hongyu He and Xueren Zheng, "Analytical model of undoped polycrystalline Silicon thin-film transistors consistent with Pao-Sah model," *IEEE Trans. on Electron Devices*, vol. 58, no. 4, pp. 1102–1107, April 2011.
45. A. O. Conde, F. J. Sanchez, J. Muci, S. Malobabic, and J. Liou, "A review of core compact models for undoped double-gate SOI MOSFETs," *IEEE Trans. on Electron Devices*, vol. 54, no. 1, pp. 131–140, Jan. 2007.
46. A. Tomkins, P. Garcia, and S. P. Voinigescu, "A 94 GHz SPST switch in 65 nm Bulk CMOS," *Proc. of Compound Semiconductor Integrated Circuits Symposium*, CA, USA, 12–15 Oct. 2008, pp. 1–4.
47. Viktor Sverdlov, "Scaling, power consumption, and mobility enhancement techniques," *Computational Microelectronics*, vol. 1, no. 1, pp. 5–22, 2011.
48. S. M. Sze, *Semiconductor devices: physics and technology*, 2nd Edition, Tata McGraw Hill, 2004.
49. Ashwani K. Rana, Narottam Chand and Vinod Kapoor, "Gate leakage behavior of source/drain-to-gate non-overlapped MOSFET structure," *J. of Computational Electronics*, vol. 10, no. 1–2, pp. 222–228, June 2011.
50. Xing Zhou, Guojun Zhu, Guan Huei See, Karthik Chandrasekaran, and Siau Ben Chiah, "Unification of MOS compact models with the unified regional modeling approach," *J. of Computational Electronics*, vol. 10, no. 1–2, pp. 121–135, 2011.
51. Hesham Hamed, Savas Kaya, Janusz and A. Starzyk, "Use of nano-scale double-gate MOSFETs in low-power tunable current mode analog circuits," *J. of Analog Integrated Circuits, Signal Processing*, vol. 54, no. 3, pp. 211–217, 2008.
52. Shyam Parthasarathy, Amit Trivedi, Saurabh Sirohi, Robert Groves, Michael Olsen, Yogesh Chauhan, Michael Carroll, Dan Kerr, Ali Tombak, and Phil Mason, "RF SOI switch FET design and modeling tradeoffs for GSM applications," *Proc. of 23rd Int. Con. on VLSI Design*, India, 3–7 Jan. 2010, pp. 194–199.
53. M. Fragopoulou, S. Siskos, M. Manolopoulou, M. Zamani, and G. Sarrabayrouse, "Thermal neutron dosimetry using MOSFET dosemeters," *J. of Radiation Measurement*, vol. 44, no. 9–10, pp. 1006–1008, Oct.-Nov. 2009.

Chapter 7
Testing of MOSFETs Surfaces Using Image Acquisition

7.1 Introduction

The image processing is frequently used in the systems for monitoring and controlling of the objects to support in an effective management of their resources and safety. The practical systems for monitoring the rectangular objects, like double-gate (DG) MOSFET, and cylindrical objects like cylindrical surrounding double-gate (CSDG) MOSFET, which requires various vision sensors, recording images that have to be transmitted to and processed in the central processing unit [1]. One of the most challenging problems in such cases is the effective transmission and processing of huge amount of image data. To avoid overloading of transmission channels and central unit, various already existing algorithms are frequently performed at the sensors by an integrated low-level image processor. As a result, the rough image data generated by the sensor can be compressed or replaced by useful information extracted from the images. This approach significantly improves the overall efficiency and the cost of the system. A complete vision chip consisting of a photodetector array, which is effectively implemented on DG MOSFET and CSDG MOSFET, is formed on the rectangular and cylindrical substrate, respectively [2].

The earlier solutions of integrated vision chips have been mostly dedicated to a specific image algorithm and could not be reconfigured. The next generations of programmable vision chips designed in single instruction multiple data (SIMD) architectures or multiple instruction multiple data (MIMD) architectures are able to perform several image algorithms [3–6]. The latest chips have fully programmable architectures with a parallel analog data processing, which significantly reduces the time required for an image processing [7, 8].

However, most of the reported vision chips realize the convolution algorithms based on a reduced kernel, where only four neighboring pixels (the top, bottom, left, and right) from the full 3 × 3 kernel are taken into calculations. The consequence of omitting the diagonal pixels is degradation of the resulting image, and some of the reported implementations [9] have speed limitation resulting from the sequential manner of the instruction processing, which means that a typical convolution algorithm requires several clock cycles to calculate the single data. In a recently emerging 3-D integrated technology, several silicon-on-insulators (SOI) with different circuits can be vertically stacked and interconnected through the 3-D bias. This high-intensity integration strategy allows a vision sensor to separate photosensitive devices with processing circuits on different tiers to achieve a high fill factor. Earlier reported literatures [10, 11] have demonstrated the image sensors designed and fabricated using this process. *Blakiewicz* [12] has proposed a matrix multiplier for an integrated low-power and low-cost image sensor design as a better alternative to known massively parallel architectures [13].

We have analyzed the image acquisition of the DG MOSFET and CSDG MOSFET for the purpose of RF switch for the advanced wireless telecommunication systems. The proposed model is emphasized on the basics of the single image sensor for 2-D images of 3-D devices, so that we can obtain a required device parameter.

The DG MOSFET and CSDG MOSFET structure utilizes an undoped body because the undoped MOSFET can avoid the dopant fluctuation effect, which contributes to the variation of the threshold voltage and drive current. So this DG MOSFET and CSDG MOSFET [14] can be used for the purpose of double-pole four-throw (DP4T) RF CMOS switch design [15–17]. Also the undoped body can enhance the carrier mobility owing to the absence of depletion charges, which significantly contributes to the effective electric field and degrades the mobility [18–20].

Here we have proposed a model for image acquisition [21] by which we validate the surface smoothness of the DG MOSFET and CSDG MOSFET. If it is smooth, then we proceed for the further testing; otherwise, the device fails. The design of the proposed model has been studied to understand the effect of device geometry useful when working as a switch. Each steps of the model are discussed separately for the purpose of clarity of presentation and understanding the process for the devices (DG MOSFET and CSDG MOSFET).

7.2 Proposed Model for the Image Acquisition of MOSFETs

Recently, the CMOS image sensors are extensively used in the commercial and scientific applications. CMOS standard processes, which are developed for the digital and mixed-signal applications, are really attractive because of their low power consumption and applicability for on-chip signal processing. Various processes have been explored to improve the image sensor performance to a very high level, and performance has been significantly enhanced with the use of CMOS image sensor processes [22]. In addition to this, the use of aggressive technologies and small MOS transistors (gate area of 1 nm^2) in the pixel are required in order to

7.2 Proposed Model for the Image Acquisition of MOSFETs   167

maximize the pixel photosensitive area. This leads to an increase of MOS transistor low-frequency noise impact [23–27]. In the digital image, pixel is a physical point in a raster image which is the smallest addressable element in a display device. In other words, we can say that the pixel is the smallest controllable element of a picture represented on the screen. The address of a pixel corresponds to its physical coordinates. However, each pixel is a sample of an original image, so large number of pixel jointly provides more accurate representations of the original structure.

To observe the parameters of the devices, we have proposed a model using image acquisition process as shown in Fig. 7.1. The details of the process steps performed in the flow graph are as follows:

7.2.1 Preprocessing

First, we put the device on the base of the image sensor to capture the image using the conveyor moving belt. This process can be performed by using the image sensor. The speed of the conveyor belt is adjusted in such a manner that the image sensor can sense the structure of the test device. Here we have to be careful about the position of the test devices on the belt which means the devices should not be overlapped to each other on the conveyor belt; otherwise, the image will be ambiguous.

7.2.2 Image Sensor

For the proposed model, we convert the DG MOSFET and CSDG MOSFET into the images. For this purpose, we have to generate a 2-D image of the device which means sensor should have displacement in X and Y direction between the sensor and the area which have to be imaged. A negative film is mounted on the MOSFETs, and a mechanical moment of MOSFET provides the displacement in one direction. A sensor is mounted on the lead that provides motion in perpendicular direction [28–30]. The preprocessing of images prior to image classification and change detection is essential, which generally comprises a series of sequential operations, including atmospheric correction or normalization, image registration, geometric correction, and masking. The geometric rectification of the imagery resamples or changes the pixel grid to fit that of a map projection or another reference image. This becomes especially important when scene-to-scene comparisons of the individual pixels in applications such as change detection are being sought. Before the creation of the minimum images, preprocessing must occur. The preprocessing procedure consists of collection, downloading, unzipping twice, executing the preprocessing algorithms through the softwares [31], checking the final preprocessed images, and executing the patch procedure.

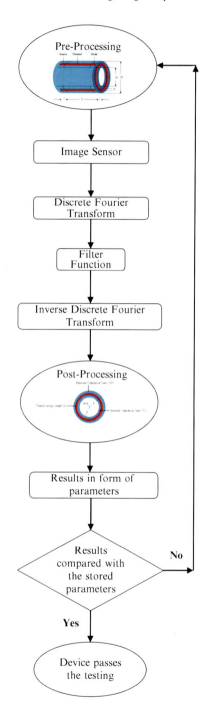

Fig. 7.1 Flow chart of a device testing using Image Acquisition

7.2 Proposed Model for the Image Acquisition of MOSFETs

The preprocessing steps of a remotely sensed image generally performed before the postprocessing enhancement, extraction, and analysis of information from the image. Typically, it will be the data provider who will preprocess the image data before delivery of the data to the customer or user. However, the preprocessing of image data often will include radiometric correction and geometric correction.

7.2.3 Discrete Fourier Transform

With the image obtained from the above image sensor process, there may be some reduction in the property of image, for example, contrast, brightness, and color. To enhance the property of image, we use the process of discrete Fourier transform (DFT) because this transform is an important tool in the area of digital signal processing. The DFT is an equivalent of the Fourier transform for discrete data. The DFT converts a finite list of equally spaced samples of a function into the list of coefficients of a finite combination of complex sinusoids, ordered by their frequencies, that has those same sample values. It can be said to convert the sampled function from its original domain to the frequency domain. In image processing, the samples can be the values of pixels along a row or column of a raster image. The DFT of N data elements $L = [L_1, \ldots, L_N]$ is defined as the list $F = [F_1, \ldots, F_N]$, such that:

$$F_k = \sum_{j=1}^{j=N} L_j e^{-2\pi i (j-1)(k-1)/N} \tag{7.1}$$

where $k = 1, 2, 3, \ldots, N$. Since the DFT requires intensive computation, so there is fast DFT processors to meet real-time signal processing requirements [32].

7.2.4 Filter Function

We have passed the above DFT image through a filter which has the filter function $H(u, v)$. This filter function is set by the requirement for the image. However, smoothing of the image is achieved in the frequency domain by dropping out the high-frequency components which means using the low-pass filter function. Sharpening, enhancing, or detecting the edges of the images is associated with high frequency, a component which is achieved with the application of high-pass filter. This filter function can be used as:

$$Y(u, v) = H(u, v) \cdot X(u, v) \tag{7.2}$$

where X is the input function and Y is the output function of the filter.

7.2.5 Inverse Discrete Fourier Transform

Finally, we compute the inverse DFT of the signal obtained from the $F(u, v)$ multiplied by a filter function $H(u, v)$. So, we obtained the enhanced image.

7.2.6 Postprocessing

However, before analyzing the image, some degree of preprocessing is necessary to correct any distortion inherent in the images due to the characteristics of imaging system and conditions. Commonly used preprocessing procedures include the radiometric correction and geometric correction. Once preprocessing is completed, the image properties can be enhanced to improve the visual appearance of the objects on the image. Generally used image enhancement techniques include image reduction, image magnification, transect extraction, contrast adjustments, band ratioing, spatial filtering, Fourier transformations, principal components analysis, and texture transformation [33]. These are used to extract useful information that assists in image interpretation. For visual image interpretation and digital image processing, the availability of secondary data and knowledge of the analyst are extremely important. The visual interpretation can be performed using various viewing and interpretation devices. However, most commonly used elements of visual analysis are tone, color, size, shape, texture, pattern, height, shadow, site, and association of the object under investigation. The digital image processing relies primarily on the radiance of image picture elements (pixels) for each band. The radiance is then translated into digital numbers or grayscale intensity, for example, from 0 (lowest intensity, or black) to 255 (highest intensity, or white). Digital numbers for a specific band indicates the intensity of the radiance at that wavelength.

7.2.7 Image Enhancement

It is a process that changes the pixel's intensity of the input image, so that the output image contains all the property of the device under test (DUT) and interpretability and perception of information contained in the image enhanced. The image enhancement is used to provide a better input for other automated image processing systems. One of the commonly used image enhancement methods is histogram equalization [25, 26]. This equalization remaps the gray level of an image based on the probability distribution of the input gray levels. It flattens and stretches the dynamic range of the image's histogram. So this results in an overall contrast enhancement by changing the intensity level of the pixels based on the intensity distribution of the input image. The subregion's histogram equalization partitions the image based on the smoothed intensity values, which have been obtained by

7.3 Image Analysis 171

convolving the input image with a Gaussian filter. With this, the transformation function used by histogram equalization is not based on the intensity of the pixels only, but the intensity values of the neighboring pixels are also taken into the consideration [34, 35].

7.3 Image Analysis

However, after obtaining the parameter from the postprocessing, we compared the parameters with the already stored parameter, which is required by the industry or customer for the further device applications. If comparison of the parameter fails, then the process will be repeated else it passes the parameter comparison, so we got the pass device.

On the way of the above processing, we can get the image of a device of various types as shown in Figs. 7.2 and 7.3. In these figures, we can compare the structure of the device. If the image obtained from the postprocessing stage is different at any

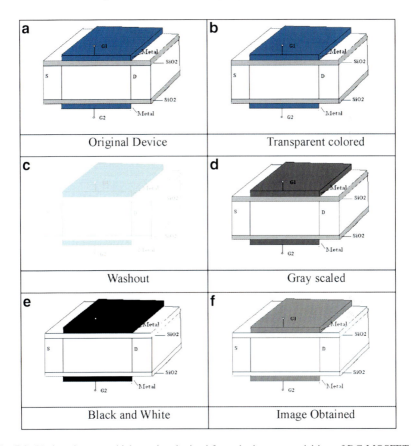

Fig. 7.2 Various images which can be obtained from the image acquisition of DG MOSFET

172   7 Testing of MOSFETs Surfaces Using Image Acquisition

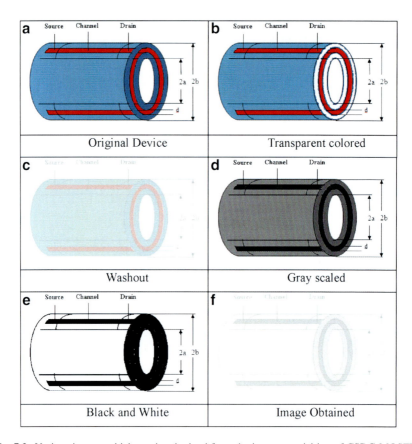

Fig. 7.3 Various images which can be obtained from the image acquisition of CSDG MOSFET

stage, we can remove the device. So the further design can be improved for the suitable structure of the device as to be cylindrical and flat at the surfaces in this particular stage. So we can improve the device structure using the image acquisition phenomenon. As we take the rectangular device (DG MOSFET) and cylindrical device (CSDG MOSFET), this process is also suitable for the other structures such as omega, FinFET, GAA etc.

7.4 Conclusion

In this chapter, we have analyzed the image acquisition of the DG MOSFET and CSDG MOSFETs for the purpose of RF switch for the application in the advanced wireless telecommunication systems. The proposed model emphasized on the basics of the single image sensor for 2-D images of 3-D devices, so that we can obtain the desired device parameter. However, using this technique we can validate the basics of the circuit elements parameter required for the radio-frequency subsystems of the integrated circuits such as capacitances, resistances, oxide thickness, resistance of polysilicon, and number of bulk capacitors [36, 37]. The model can be easily introduced in the circuit simulators.

However, beyond the proposed model in this book, it has some potential challenges such as the accuracy of the device contact which is a challenging process. Here we apparently restrict the analysis of the spherical devices because it requires a number of image sensors to take the image of device from various angles. As a consequence, the comparisons have been made in regard to the available modes of the image sensor.

References

1. P. Dutkiewicz, M. Kieczewski, K. Kozowski, and D. Pazderski, "Vision localization system for mobile robot with velocities and acceleration estimator," *Bulletin of the Polish Academy of Sciences and Technical Sciences*, vol. 58, no. 1, pp. 29–41, Dec. 2010.
2. W. Jendernalik, J. Jakusz, G. Blakiewicz, R. Piotrowski, and S. Szczepanski, "CMOS realisation of analogue processor for early vision processing," *Bulletin of the Polish Academy of Sciences and Technical Sciences*, vol. 59, no. 2, pp. 141-147, Aug. 2011.
3. R. E. Cummings, Z. K. Kalayjian, and D. Cai, "A programmable focal plane MIMD image processor chip," *IEEE J. Solid State Circuits*, vol. 36, no. 1, pp. 64–73, Jan. 2001.
4. J. Schemmel, K. Meier, and M. Loose, "A scalable switched capacitor realization of the resistive fuse network," *Analog Integrated Circuits and Signal Processing*, vol. 32, no. 2, pp. 135–148, Aug. 2002.
5. A. Dupret, J. O. Klein, and A. Nshare, "A DSP-like analog processing unit for smart image sensors," *Int. J. Circuit Theory and Application*, vol. 30, no. 6, pp. 595–609, 2002.
6. D. A. Martin, H. S. Lee, and I. Masaki, "A mixed signal array processor with early vision applications," *IEEE J. Solid State Circuits*, vol. 33, no. 3, pp. 497–502, March 1998.
7. P. Dudek and P. J. Hicks, "An analogue SIMD focal plane processor array," *Proc. of IEEE Int. Symp. on Circuits and Systems*, Sydney, Australia, 6-9 May 2001, pp. 490–493.
8. P. Dudek and P. J. Hicks, "A general purpose processor per-pixel analog SIMD vision chip," *IEEE Trans. Circuits and Systems*, vol. 52, no. 1, pp. 13–20, Jan. 2005.
9. P. Dudek, A. Lopich, and V. Gruev, "A pixel parallel cellular processor array in a stacked three layer 3D silicon-on-insulator technology," *Proc. of Eur. Conf. on Circuit Theory and Design*, Turkey, Antalya, 23–27 Aug. 2009, pp. 193–196.
10. V. Suntharalingam, R. Berger, J. Bums, and C. Chen, "Megapixel CMOS image sensor fabricated in three dimensional integrated circuit technology," *Proc. of IEEE Solid State Circuits Conf.*, Pennsylvania, USA, 10 Feb. 2005, pp. 356-357.
11. E. Culurciello and P. Weerakoon, "Three dimensional photo detectors in 3D silicon-on-insulator technology," *IEEE Electron Device Letters*, vol. 28, pp.117-119, 2007.
12. G. Blakiewicz, "Analog multiplier for a low-power integrated image sensor," 16th *Int. Conf. on Mixed Design of Integrated Circuits and Systems*, Poland, 25-27 June 2009, pp. 226-229.
13. P. Dollfus, "Sensitivity of single and double-gate MOS architectures to residual discrete dopant distribution in the channel," *J. of Computational Electronics*, vol. 5, no. 2-3, pp. 119-123, July 2006.

14. Viranjay M. Srivastava, K. S. Yadav, and G. Singh, "Design and performance analysis of cylindrical surrounding double-gate MOSFET for RF switch," *Microelectronics Journal*, vol. 42, no. 10, pp. 1124-1135, Oct. 2011.
15. Viranjay M. Srivastava, K. S. Yadav, and G. Singh, "Analysis of double gate CMOS for DP4T RF switch design at 45-nm technology," *J. of Computational Electronics*, vol. 10, no. 1-2, pp. 229-240, June 2011.
16. M. Cheralathan, Antonio Cerdeira, and Benjamin Iniguez, "Compact model for long-channel cylindrical surrounding-gate MOSFETs valid from low to high doping concentrations," *Solid State Electronics*, vol. 55, no. 1, pp. 13-18, Jan. 2011.
17. S. Kolberg, H. Borli, and T. A. Fjeldly, "Modeling, verification and comparison of short-channel double gate and gate-all-around MOSFETs," *Mathematics and Computers in Simulation*, vol. 79, no. 4, pp. 1107-1115, Dec. 2008.
18. M. Reyboz, P. Martin, T. Poiroux, and O. Rozeau, "Continuous model for independent double gate MOSFET," *Solid State Electronics*, vol. 53, no. 5, pp. 504-513, May 2009.
19. Riza Tamer and Kausik Roy, "Analysis of options in double-gate MOS technology: A circuit perspective," *IEEE Trans. on Electron Devices*, vol. 54, no. 12, pp. 3361–3368, Dec. 2007.
20. Antonio Cerdeira, Benjamin Iniguez, and Magali Estrada, "Compact model for short channel symmetric doped double-gate MOSFETs," *Solid State Electronics*, vol. 52, no. 7, pp. 1064-1070, July 2008.
21. A. G. Andreou, R. C. Meitzler, K. Strohbehn, and K. A. Boahen, "Analog VLSI neuromorphic image acquisition and pre-processing systems," *Neural Networks*, vol. 8, no. 7–8, pp. 1323-1347, 1995.
22. M. Furumiya, "High sensitivity and no-crosstalk pixel technology for embedded CMOS image sensor," *IEEE Trans. Electron Devices*, vol. 48, no. 10, pp. 2221–2227, Oct. 2001.
23. P. Gonthier, E. Havard, and P. Magnan, "Custom transistor layout design techniques for random telegraph signal noise reduction in CMOS image sensors," *Electronics Letters*, vol. 46, no. 19, pp. 1323-1324, Sept. 2010
24. Maria Petrou, and Panagiota Bosdogianni, *Image processing: The fundamental*, John Wiley & Sons Ltd, New York, 2000.
25. Rafael C. Gonzalez, and Richard E. Woods, *Digital Image Processing*, 2nd Edition, Prentice-Hall of India, New Delhi, 2002.
26. Viranjay M. Srivastava, K. S. Yadav, and G. Singh, "Explicit model of cylindrical surrounding double-gate MOSFETs," *WSEAS Trans. on Circuits and Systems*, vol. 12, no. 3, pp. 81–90, March 2013.
27. C. Rubat Du Merac, P. Jutier, J. Laurent, and B. Courtois, "A new domain for image analysis: VLSI circuits testing, with Romuald, specialized in parallel image processing," *Pattern Recognition Letters*, vol. 1, no. 5–6, pp. 347-352, July 1983.
28. A. Gamal and H. Eltoukhy, "CMOS image sensors," *IEEE Circuits and Devices Magazine*, vol. 21, no. 3, pp. 6–20, March 2005.
29. Jiaming Tan, B. Buttgen, and A. Theuwissen, "Analyzing the radiation degradation of 4-transistor deep submicron technology CMOS image sensors," *IEEE Sensors J.*, vol. 12, no. 6, pp. 2278-2286, June 2012.
30. H. D. Cheng, Y. Y. Tang, and C. Y. Suen, "Parallel image transformation and its VLSI implementation," *Pattern Recognition*, vol. 23, no. 10, pp. 1113-1129, 1990.
31. International Technology Roadmap for Semiconductors-2010, www.public.itrs.net
32. Ching Hsien Chang, Chin Liang Wang, and Yu Tai Chang, "Efficient VLSI architectures for fast computation of the discrete Fourier transform and its inverse," *IEEE Trans. on Signal Processing*, vol. 48, no. 11, pp. 3206-3216, Nov. 2000.
33. L. D. Van, Yuan Chu Yu, Chun Ming Huang, and Chin Teng Lin, "Low computation cycle and high speed recursive DFT/IDFT: VLSI algorithm and architecture," *Proc. IEEE Workshop on Signal Processing Systems Design and Implementation*, Athens, Greece, 2–4 Nov. 2005, pp. 579- 584.

References

34. Haidi Ibrahim and Nicholas Kong, "Image sharpening using sub-regions histogram equalization," *IEEE Trans. on Consumer Electronics,* vol. 55, no. 2, pp. 891-895, May 2009.
35. A. G. Corry, D. K. Arvind, G. S. Connolly, R. R. Korya, and I. N. Parker, "Image processing with VLSI," *Microprocessors and Microsystems,* vol. 7, no. 10, pp. 482-486, Dec. 1983.
36. Viranjay M. Srivastava, K. S. Yadav, and G. Singh, "DP4T RF CMOS switch: A better option to replace SPDT switch and DPDT switch," *Recent Patents on Electrical and Electronic Engineering,* vol. 5, no. 3, pp. 244–248, Oct. 2012.
37. Viranjay M. Srivastava, K. S. Yadav, and G. Singh, "Drain current and noise model of cylindrical surrounding double-gate MOSFET for RF switch," *Procedia Engineering,* vol. 38, pp. 517–521, April 2012.

Chapter 8
Conclusions and Future Scope

8.1 Conclusions

After modeling of the symmetrical double-gate (DG) MOSFET, we have drawn the layout and simulate with the parameters available in the MOSFET. It includes the basics of the circuit elements parameter required for the radio-frequency subsystems of the integrated circuits such as drain current, output voltage, threshold voltage, capacitances, resistances at switch ON-state condition, oxide thickness, resistance of polysilicon, number of bulk capacitors, and power or voltage gain. For the purpose of radio-frequency (RF)/microwave switch, we have explored the methods to minimize the control voltage, resistance for the switching condition, and the capacitances for isolation. However, double-pole four-throw (DP4T) CMOS inverter switch has been designed with low insertion loss and low control voltage. The DP4T RF CMOS switch has been designed using the independently controlled double-gate, which has been discussed in detail, and the impact on the power consumption with respect to ON-state resistance and current, propagation delay, leakage behavior, as well as area of the device is presented. It shows that the numbers of transistor are reduced with the application of DG MOSFET and also the area can be significantly reduced for logic gates; therefore, the logic density per area increases. The favorable condition for low-power circuit is that where both transistor gates are on the same potential contribution even a reduced amount of the leakage current. The proposed DP4T RF CMOS switch design with double-gate transistors modifies the conventional analog switch circuit design to operate with digital signals to achieve isolation buffering for bidirectional signals. High-density packing of multiple buffer switches operating under single enable control in a single package can also be achieved.

In the DG MOSFET, the bulk voltage is 0, so we achieved the higher drain current. We have reported the attenuation of 0.005–0.016 dB for 45-nm technology compared to the attenuation of 0.020–0.070 dB for 0.8-μm technology. The off-isolation and switching speed are significantly improved in the discussed DP4T DG RF CMOS switch over the already existing CMOS switch.

Moreover, the flat-band capacitance and power dissipation become half, and threshold voltage as well as flat-band voltage is reduced as flat-band capacitance becomes half for the discussed DP4T DG RF CMOS switch. The half power dissipation has been discussed for the proposed DP4T DG RF CMOS switch, and the results were compared with the already existing CMOS switches. The ultrathin body silicon-on-insulator (SOI) FETs suggests a very thin silicon body to achieve better control of the channel by the gate and, hence, reduces the leakage and short-channel effects. With the use of the intrinsic or lightly doped body, in the DG MOSFET, reduction in the threshold voltage occurs due to the random dopant fluctuations, which enhances the mobility of the careers in the channel region, and therefore increment in ON-state current occurs. By using the designed capacitive model of the DP4T DG RF CMOS switch, the equivalent circuit of this model for the switch and simulated S-parameters are presented. For the purpose of RF switch, we have achieved the process to minimize the ON-state resistance and maximize the parasitic capacitance and minimize the control voltage to control the isolation and switching speed for the switch-ON condition with DP4T DG RF CMOS switch. Since the operating frequencies of the RF switches are in the range of few MHz to 60 GHz, therefore, it is useful for modern broadband wireless communication systems. The proposed DP4T DG RF CMOS switch results to the peak output currents around 0.387 mA and switching speed of 36 ps. A device structure with a double-gate contact shows a significant improvement in currents and switching speed as compared to the single-gate structure.

For increasing the gate voltage, the drain current increases; hence, the contact resistance decreases which increases the cutoff frequency. Therefore, for the purpose of RF switch, control voltage should be low and then current flow will be less. In the terms of contact resistance, it increases with the increase in the number of gate fingers. So, in the application of RF switch, we have tried to increase the gate finger. From the simulated result of DG MOSFET, we conclude that (for the DG MOSFET, bulk voltage is 0) highest drain current can be easily achieved by using higher gate fingers. Higher gate fingers, enhances the mobility of the charge careers in the channel region as compared to the lower gate fingers due to intrinsic or light doping, and therefore, increment in drain current occurs at higher gate fingers.

We have analyzed the designing parameters of a cylindrical surrounding double-gate (CSDG) MOSFET. The CSDG MOSFET device is operated in the microwave frequency regime of the spectrum. For the purpose of RF or microwave switch, we achieved the process to minimize control voltage, capacitances for isolation, and the resistance for the switching condition and increased energy storage of a device. From the discussions in previous chapters, we have achieved a better CSDG MOSFET as compared with cylindrical surrounding single-gate (CSSG) MOSFET.

A CSDG MOSFET has been designed, and we have simulated the parameters available in this model, by using the available tools. It includes the basics of the circuit elements parameter required for the radio-frequency subsystems of the integrated circuits such as drain current, output voltage, threshold voltage, capacitances, resistances at switch-ON condition, oxide thickness, resistance of polysilicon, energy stored, cross talk, number of bulk capacitors, and power or voltage gain. For the RF/microwave switch, we have achieved the process to minimize the control voltage, capacitances for isolation, and the resistance for the switching condition and increased energy storage of a device. The mobile charge density is calculated by using the analytical expressions obtained from modeling the surface potential as well as the difference of potentials at the surface and at the center of the silicon-doped layer. The analytical expressions for the charge characteristics are presented as the function of silicon layer impurity concentration, gate dielectric, and silicon layer thickness with the variable mobility.

We extended this work with a replacement of SiO_2 with HfO_2 as a high dielectric material. Hafnium dioxide (HfO_2) is a high dielectric, low-absorption material usable for coatings in UV (~250 nm) to IR (~10 μm) regions; its adhesion is superb to metals such as aluminum and silver. In the presented work, CMOS with pure SiO_2 is adopted for DP4T DG MOSFET switch to detect the signals to/from transceivers, and the performance of this DP4T DG RF CMOS switch at 45-nm technology was demonstrated. The performance of HfO_2 for switch as effective R_{ON}, attenuation, flat-band capacitance, average dynamic power, and working efficiency at high temperature have been investigated. In the DP4T switch, effective resistance R_{ON} should be less, and it is inversely proportional to intrinsic transconductance (which is higher in HfO_2 compare to SiO_2), so it is better to use HfO_2 for a DP4T switch. Also, the attenuation is directly proportional to R_{ON}, so it can be improved by using the HfO_2.

After designing of the various devices as discussed in the previous chapters, we have proposed a model to observe the smoothness of the device surfaces. We have analyzed the image acquisition of the DG MOSFET and CSDG MOSFETs. The proposed model emphasized on the basics of the single image sensor for two-dimensional images of three-dimensional devices, so that we can obtained a satisfactory device parameter. The model can be easily introduced in circuit simulators. Beyond the proposed model in this book, it has some potential challenges such as the accuracy of the device contact which is a challenging process. Here we apparently restrict the analysis of the spherical devices because it requires a number of image sensors to take the image of device from various angles. As a consequence, the comparisons have been made in regard to the available modes of the image sensor.

8.2 Future Scope

However, beyond the analyzed parameters in this work, the proposed CSDG MOSFET has potential challenges such as the fabrication of this kind of devices by a tricky process. In the presented work, we apparently restrict our comparative analysis of the different devices to an applied high drain biasing [1]. As a consequence, the comparisons have been made, particularly, in regard to the RF performance and saturation regime, the most important of which is transit frequency,

f_T (subthreshold swing and OFF-state current), which is not covered in this work that is imperative for the wireless applications [2]. Another decisive factor for the microwave application is the hot carrier effect, which affects the device degradation for high-power and high-frequency applications, which is a very interesting problem of research [3]. In general, the physical oxide thickness is determined by ellipsometer, but its accuracy cannot be guaranteed for those of very thin oxides.

As a consequence of the power-supply voltage being reduced much less proportionally to the channel length, the lateral electric field will be increased in the device. The carriers which move from the source to the drain in such a turned-ON transistor can get enough energy to cause impact ionization that generates electron–hole pairs in silicon and surmount the interfacial energy barrier. The energy of the hot carriers depends mainly on the electric field in the pinch-off region. The carriers injected into a gate dielectric induce device degradation such as threshold voltage (V_{th}) shift due to occupied traps in the oxide and reduced drain current (I_{ds}). The hot carriers can also generate traps at the silicon–oxide interface leading to subthreshold swing deterioration and stress-induced drain leakage [4, 5]. Therefore, the hot carrier injection degradation significantly reduces the transistor lifetime. The n-type MOSFET is more sensitive to hot carrier injection than the p-type MOSFET transistor, since electrons become hotter than holes due to their higher mobility and energy barrier is lower for electrons compared to holes at the interface [6, 7]. However, to solve this issue, drain engineering is used to alleviate the peak of the lateral electric field located close to the drain edge by modifying the drain doping profile through introduction of source/drain extension implants by a lower dose [8, 9].

In general, one may use a combination of two or more metals on a single Si-substrate to achieve the work function requirements discussed in the previous chapters. It must be noted that the integration of multiple metals on a single wafer poses significant process integration challenges, and it would be highly desirable to develop a single metal tunable work function gate CMOS process. Such a process would allow for minimal process complexity and would be relatively easy to integrate on a single Si-substrate [10]. Probably the most intuitive approach to developing a metal gate CMOS process involves the use of two metals, on serving as the n-type MOS gate and the other as the p-type MOS gate. Process integration is however not very straightforward since two film deposition steps and at least two etch steps would be involved [11]. The general process involves the deposition of one metal over the entire substrate after active area definition and gate dielectric deposition. Following the first deposition, a well lithography is performed to expose both n-well or p-well regions, and the exposed metal is removed using a wet etch chemistry. This can be extending in future works.

However, another approach is using high-resolution transmission electron microscopy analysis. This method is more accurate but still suffers from high cost and low throughput. In addition, the thickness measured with these methods is physical thickness. It cannot be employed to determine the equivalent oxide thickness of the high dielectric constant (high-k) materials, proposed for future

ULSI CMOS applications, because dielectric constants of these materials are different to that of oxide [12–17].

We have analyzed the image acquisition model for the DG MOSFET and CSDG MOSFETs as to observe the smoothness of the device surfaces. Beyond the proposed model in this book, it has some potential challenges such as the accuracy of the device contact [18]. Here we apparently restrict the analysis of the spherical devices because it requires a number of image sensors to take the image of device from various angles [19–22]. As a consequence, the comparisons can be made in regard to the various modes of the image sensor.

References

1. Ji Wu and Gaofeng Wang, "Cost evaluation on reuse of generic network service dies in three-dimensional integrated circuits," *Microelectronics Journal*, vol. 44, no. 2, pp. 152–162, Feb. 2013.
2. Shih Chang Tsai, San Lein Wu, Bo Chin Wang, Shoou Jinn Chang, Che Hua Hsu, and J. F. Chen, "Low-frequency noise characteristics for various ZrO_2-added HfO_2-based 28-nm high-k /metal-gate n-MOSFETs," *IEEE Electron Device Letters*, vol. 34, no. 7, pp. 834–836, July 2013.
3. Juun Yu Tsai, Ting Chang Wen Hung Lo, Ching En Chen, Szu Han Ho, Hua Mao Chen, Ya Hsiang Tai, and Cheng Tung Huang, "Abnormal sub-threshold swing degradation under dynamic hot carrier stress in HfO_2/TiN n-channel metal-oxide-semiconductor field-effect-transistors," *Applied Physics Letters*, vol. 103, no. 2, pp. 106–108, July 2013.
4. S. Garduno, A. Cerdeira, M. Estrada, J. Alvarado, V. Kilchytska, and D. Flandre, "Improved modeling of gate leakage currents for Fin–shaped field–effect transistors," *J. of Applied Physics*, vol. 113, no. 12, pp. 507–509, March 2013.
5. S. Das and S. Kundu, "Simulation to study the effect of oxide thickness and high-k dielectric on drain induced barrier lowering in n-type MOSFET," *IEEE Trans. on Nanotechnology*, 2014, in press.
6. J. W. Liu, M. Y. Liao, M. Imura, H. Oosato, E. Watanabe, A. Tanaka, and Y. Koide, "Interfacial band configuration and electrical properties of $LaAlO_3/Al_2O_3$/hydrogenated-diamond metal-oxide-semiconductor field effect transistors," *J. of Applied Physics*, vol. 114, no. 8, pp. 108–114, Aug. 2013.
7. X. Gong, B. Liu, and Y. Yeo, "Gate Stack Reliability of MOSFETs with High Mobility Channel Materials: Bias Temperature Instability," *IEEE Trans. Device and Materials Reliability*, 2014, in press.
8. Yuya Minoura, Atsushi Kasuya, Takuji Hosoi, Takayoshi Shimura, Heiji Watanabe, "Design and control of Ge-based metal-oxide-semiconductor interfaces for high-mobility field-effect transistors with ultrathin oxynitride gate dielectrics," *Applied Physics Letters*, vol. 103, no. 3, pp. 502–506, July 2013.
9. Udit Monga, Dag M. Nilsen, Tor A. Fjeldly, "Modeling of electrostatics and drain current in nanoscale quadruple-gate MOSFET using conformal mapping techniques," *Microelectronics Journal*, vol. 44, no. 1, pp. 3–6, Jan. 2013.
10. S. Venugopalan, M. A. Karim, S. Salahuddin, A. M. Niknejad, and C. C. Hu, "Phenomenological compact model for QM charge centroid in multigate FETs," *IEEE Trans. on Electron Devices*, vol. 60, no. 4, pp. 1480–1484, April 2013.
11. P. Suveetha Dhanaselvam and N. B. Balamurugan, "Analytical approach of a nanoscale triple-material surrounding gate (TMSG) MOSFETs for reduced short-channel effects," *Microelectronics Journal*, vol. 44, no. 5, pp. 400–404, May 2013.

12. Zhong Wang, Jason Ralph, Naser Sedghi, and Steve Hall, "Bound states within the notch of the HfO$_2$/GeO$_2$/Ge stack," *J. of Vacuum Science and Technology B: Microelectronics and Nanometer Structures*, vol. 31, no. 2, pp. 209–214, March 2013.
13. X. Sun, O. I. Saadat, K. S. Chang, T. Palacios, S. Cui, and T. P. Ma, "Study of gate oxide traps in HfO$_2$/AlGaN/GaN metal-oxide-semiconductor high-electron-mobility transistors by use of ac transconductance method," *Applied Physics Letters*, vol. 102, no. 10, pp. 504–507, March 2013.
14. K. Manabe, K. Watanabe, H. Jagannathan, and V. K. Paruchuri, "Mechanism for leakage reduction by la incorporation in a HfO$_2$/SiO$_2$/Si gate stack," *IEEE Electron Device Letters*, vol. 34, no. 3, pp. 348–350, March 2013.
15. Rui Zhang, Po Chin Huang, Ju Chin Lin, N. Taoka, M. Takenaka, and S. Takagi, "High mobility Ge p- and n-MOSFETs with 0.7 nm EOT using HfO$_2$/Al$_2$O$_3$/GeO$_x$/Ge gate stacks fabricated by plasma post oxidation," *IEEE Trans. on Electron Devices*, vol. 60, no. 3, pp. 927–934, March 2013.
16. S. Knebel, S. Kupke, U. Schroeder, S. Slesazeck, T. Mikolajick, R. Agaiby, and M. Trentzsch, "Influence of frequency dependent time to breakdown on high-k/metal gate reliability," *IEEE Trans. on Electron Devices*, vol. 60, no. 7, pp. 2368–2371, July 2013.
17. Szu Ho, Ting Chang, Ying Lu, Wne Lo, and Ching Chen, "Investigation of extra traps measured by charge pumping technique in high voltage zone in p-channel metal-oxide-semiconductor field-effect transistors with HfO$_2$/metal gate stacks," *Applied Physics Letters*, vol. 102, no. 1, pp. 106–109, Jan. 2013.
18. Viranjay M. Srivastava, K. S. Yadav, and G. Singh, "DP4T RF CMOS switch: A better option to replace SPDT switch and DPDT switch," *Recent Patents on Electrical and Electronic Engineering*, vol. 5, no. 3, pp. 244–248, Oct. 2012.
19. K. Ito, C. Niclass, I. Aoyagi, H. Matsubara, M. Soga, S. Kato, M. Maeda, and M. Kagami, "System design and performance characterization of a MEMS-based laser scanning time-of-flight sensor based on a 256 x 64-pixel single-photon imager," *IEEE Photonics Journal*, vol. 5, no. 2, p. 6800114, April 2013.
20. J. Szelc and H. Rutt, "Near-field THz imaging and spectroscopy using a multiple subwavelength aperture modulator," *IEEE Transactions on Terahertz Science and Technology*, vol. 3, no. 2, pp. 165–171, March 2013.
21. Viranjay M. Srivastava and G. Singh, "Testing of cylindrical surrounding double-gate MOSFET parameters using image acquisition," *J. of Signal Processing Theory and Applications*, vol. 2, no. 1, pp. 43–53, Aug. 2013.
22. C. C. Liebe and K. Coste, "Distance measurement utilizing image-based triangulation," *IEEE Sensors Journal*, vol. 13, no. 1, pp. 234–244, Jan. 2013.

Appendix A
List of Symbols

Symbols	Explanation
1 dB	1 Decibel
2-D	Two dimensional
3-D	Three dimensional
A	Gate area
AC	Alternating current
C_{bd}	Bulk-drain junction capacitance
C_{bs}	Bulk-source junction capacitance
C_{ds}	Drain-source capacitance
C_{DG}	Capacitance of DG MOSFET
C_{dSi}	Drain to source intrinsic capacitance
C_G	Gate capacitance
C_{gb}	Gate to bulk capacitance
C_{gd}	Gate to drain capacitance
C_{gs}	Gate to source capacitance
C_{in}	Input capacitance
CMOS	Complementary metal oxide semiconductor
C_{out}	Output capacitance
C_p	Capacitance value measured with parallel equivalent circuit model
C_s	Capacitance value measured with series equivalent circuit model
DC	Direct current
DIBL	Drain-induced barrier lowering (short channel effect)
DUT	Device under test
E_C	Conduction band
E_g	Band gap of silicon, 1.12 ev at 300 K
ESD	Electrostatic discharge
FD	Fully depleted
FET	Field effect transistor
F_{max}	Maximum frequency of oscillation
f_T	Cross-over frequency
GaAs	Gallium arsenide
GHz	Gigahertz

(continued)

(continued)

Symbols	Explanation
HBM	Human body model
IC	Integrated circuit
I_d	Drain current
I_{dc}	Direct current electricity
I_{ds}	Drain to source current
IF	Intermediate frequency
IL	Insertion loss
ILD	Inter layer dielectric
ISO	Isolation loss
k	Boltzmann constant
L	Channel length
LNA	Low noise amplifier
MEMS	Micro electro mechanical system
MESFET	Metal Semiconductor Field Effect Transistor
MMIC	Monolithic microwave integrated circuit
MOS	Metal oxide semiconductor
MOSFET	Metal oxide semiconductor field effect transistor
n	Concentration (density) of free electrons
N_A	Channel doping concentration of acceptors
$N_{D/S}$	Source/drain doping concentration
NF	Number of parallel devices in a multi-finger layout
n_i	Intrinsic electron density in silicon, 1.45×10^{10} cm^{-3} at 300 K
nm	Nanometer
NMOS	N type MOS transistor
NQS	Non-quasi-static
PA	Power amplifier
PD	Partially depleted
PIN	P-type-Intrinsic-N-type
q	Electron charge
$q/(kT)$	38.68 V^{-1} at 300 K
Q_B	Bulk charge
Q_{BD}	Bulk to drain junction charge
Q_{BS}	Bulk to source junction charge
Q_D	Drain charge
Q_{GB}	Gate to bulk charge
Q_I	Inversion charge
Q_{inv}	Inversion carrier sheet density
Q_S	Source charge
Q_{TH}	Inversion carrier sheet density under threshold conditions
R	Resistance
R_b	Bulk resistance
R_{bd}	Substrate resistance between substrate drain node and bulk node
R_{bs}	Substrate resistance between substrate source node and bulk node
R_d	Drain resistance
R_{dc}	Direct current resistance
R_{DG}	Resistance of DG MOSFET

(continued)

(continued)

Symbols	Explanation
R_{ds}	Parasitic drain to source resistance in parallel with the MOS channel
R_{dsb}	Substrate resistance between internal bulk node and substrate node
RF	Radio frequency
RFC	Radio frequency chokes
R_g	Gate resistance
R_{in}	Input resistance
R_{out}	Output resistance
R_p	Equivalent parallel resistance measured with parallel equivalent circuit model
R_s	Equivalent series resistance measured with series equivalent circuit model
s	Laplace term
S	Scattering parameters
SAW	Surface acoustic wave
SOI	Silicon on insulator
SOS	Silicon on sapphire
T	Absolute temperature
T/R	Transmit/receive
t_I	Thickness of the gate dielectric (insulator) layer
t_{ox}	Thickness of the gate oxide layer
t_{Si}	Thickness of silicon film (channel)
V_{bd}	Body to drain voltage
VCO	Voltage-controlled oscillator
V_{dc}	Direct current voltage
V_{dg}	Drain to gate voltage
V_{ds}	Drain to source voltage
$V_{F,eff}$	Effective front gate voltage
V_{gs}	Gate to source voltage
$V_{gs,B}$	Bottom (back) gate voltage
$V_{gs,F}$	Front gate voltage
V_H	Higher effective gate voltage
V_{in}	Input voltage
V_L	Lower effective gate voltage
V_{out}	Output voltage
V_{th}	Threshold voltage
V_T	Thermal voltage
W	Channel width
WLAN	Wireless local area networks
X	Reactance
Y	Admittance
Z	Impedance
Z_0	Reference impedance level (50 Ω in this book)
Z_{in}	Input impedance
Z_{out}	Output impedance
ε_0	Dielectric constant of vacuum, $8.854*10^{-12}$ Fm^{-1}
ε_I	Permittivity (relative dielectric constant) of gate dielectric (insulator)

(continued)

(continued)

Symbols	Explanation
ε_{Si}	Permittivity (relative dielectric constant) of silicon, 11.9
ε_{SiO2}	Relative dielectric constant of silicon dioxide (SiO_2), 3.9
λ_{Di}	Intrinsic Debye length, 48.49 μm at 300 K
φ_F	Potential difference between Fermi level and quasi-Fermi level
Φ_i	Work function of intrinsic silicon, 4.71 ev at 300 K
Φ_M	Work function of gate material
Φ_{MB}	Work function of bottom (back) gate material
Φ_{MF}	Work function of front gate material
Φ_{MS}	Work function difference between gate material and doped silicon
$\Phi_{MS,i}$	Work function difference between gate material and intrinsic silicon
ω	Angular frequency
ΔS_{11m}	Measurement error of S_{11} magnitude
ΔS_{12m}	Measurement error of S_{12} magnitude
ΔS_{21m}	Measurement error of S_{21} magnitude
ΔS_{22m}	Measurement error of S_{22} magnitude
μm	Micrometer

Appendix B
List of Definitions

1 dB compression point	The 1 dB gain compression point is the input power level in dBm at which the overall gain of amplifier is reduced by 1 dB from its maximum value
Design for manufacturability (DFM)	Design and verification methodology employed to assure that production silicon yields a suitable percentage of die, meeting the design specifications
Design for performance (DFP)	Design and verification methodology employed to assure that production silicon meets the performance objectives of the original circuit design
Design for reliability (DFR)	Design and verification methodology employed to assure that the IC performance does not substantially degrade over the anticipated lifetime of the product
Design rule check (DRC)	A check of physical layout using a foundry-determined set of rules or more complex computations to include the effect of nearby lithographic patterns
Design rules	Design rules are constraints poses by the processing line in the form of minimum allowable values for width, separation, extension and overlap. The complexity of design rules depends upon how well a process is characterized

Doped semiconductor	To improve the conductivity, impurities are added in intrinsic (pure) semiconductors. To increase the number of electrons Arsenic or Phosphorous is added into Si. It has five electrons in its outer shell. Four of these five electrons bond with adjacent silicon atoms, but the fifth electron cannot form a bond
Drain-induced barrier lowering (DIBL)	The effect of drain field reaching through the channel and lowering the source-channel energy barrier, thereby resulting in reductions in threshold and increases in output conductance with drain voltage
Fabrication steps	These steps are Wafer processing, Mask making, Photolithography, Oxidation, Diffusion, Etching, Poly-gate formation, and Metallization
Figure-of-merit (FOM)	A normalized quantity calculated based on several analog circuit performance parameters that enable a comparison of the quality of circuits having different performance parameters
Finite difference method (FDM)	The FDM subdivides the simulation domain into small discrete segments separated by nodal points. The method is based on defining unknown variable only on these nodal points assuming linear variation in between
Gain-bandwidth product (GBW)	Gain-bandwidth is defined simply as $gm = Cox$ and is used here for the purpose of determining how GBW varies, rather than a numerical value
High dielectric constant (High-K)	Term used in nanoscale technology nodes referring to gate materials that have higher dielectric constants than oxynitrde, such as hafnium and zirconium
Hot carrier injection (HCI)	Carrier injection into the channel or gate insulator produced by impact ionization near the drain end of the channel creating interface and oxide trap damage
Low dielectric-constant material	The material having lower dielectric constant than the SiO_2 used prior to nanoscale technology nodes
Low-power process (LP) device	Low-power incorporate thicker oxides than the high-performance to reduce logic power using higher thresholds and lower leakages. LP have higher saturation voltages, lower transconductance, higher output conductance, and significantly lower gain-bandwidths

Appendix B

Maximum excess overdrive supply voltage ($V_{CC,EOD}$)	The maximum drain-to-source voltage for devices with drain voltage waveforms that allow safe operation at voltages above $V_{CC,OD}$
Maximum overdrive supply voltage ($V_{CC,OD}$)	The maximum drain-to-source voltage for core devices that are not operated at high drive currents
Maximum supply voltage ($V_{CC,MAX}$)	The maximum gates to source drain to source, or source to bulk voltage for simulation purposes
Metal–insulator–metal capacitor (MIM)	It is a technology using a thin insulator between intermediate interconnect metal and an added metal to create a precise capacitor in an analog process
Minimum-gate-length feature size	Minimum gate feature size in the physical layout
Negative bias temperature instability (NBTI)	Threshold voltage instability in p-MOS devices that is dependent on temperature and device geometry
Overdrive (OD)	An operating condition permitted by design rules where some devices, such as core devices, may exceed normal operating parameters such as voltages
Power amplifier (PA)	A power amplifier is a component of transmitter frontend, used to amplify the RF signal to very high levels for transmission from the external antenna
Scaling of device	It refers to ordered reduction in dimensions of the MOSFET and other VLSI features. It allows the same decision to be made using less power and area and thus drives the electronic revolution
System-on-chip (SOC)	Technology that enables integration of all necessary electronic circuits for a complete system on a single integrated circuit
Technology based computer-aided design (TCAD)	It is used to define the physical configuration of devices within the silicon and then determines device (or simple circuit) performance using physical models of carrier propagation
Threshold voltage	The voltage on gate required to create inversion layer is called threshold voltage

Appendix C
Outcomes of the Book

1. Each gate controls one half of the device and its operation is completely independent of the other
2. Two independent MOSFETs on a single chip whose operations are independent of the other
3. Board component count and hence total cost decreases
4. The total current through the device should be the sum of the currents through the separate devices
5. The performance of the symmetrical version of the DG MOSFET is further increased by higher channel mobility compared to a bulk MOSFET
6. The average electric field in the channel is lower, which reduces interface roughness scattering according to the universal mobility model
7. One of the major advantages of using DG MOSFET is the lower leakage current and smaller subthreshold voltage
8. For single gate MOSFET, at ON condition of a transistor, increasing C_{sb} and C_{db} tends to more signal being coupled with the substrate and dissipated in the substrate resistance (R_b), so the design would like that no signal being coupled with the substrate and dissipated in the substrate resistance
9. The isolation is better in DG MOSFET compared to that in the single-gate MOSFET in terms of drain-to-source capacitance ($C_{DG} > C_{SG}$)
10. ON-state resistance is low, which shows that the current flows from source-to-drain in a MOSFET ($R_{DG} < R_{SG}$)
11. For appropriate working of a switch and to reduce the insertion loss, we have achieved the reduction in ON-state resistance with selecting transistor with large μ, increasing W/L, keeping $V_{gs} - V_{th}$ large
12. Bulk capacitors are not taken into account for the design
13. Highest drain current can be easily achieved by using the higher number of fingers
14. Reduction in the threshold voltage occurs due to the random dopant fluctuations (by the use of intrinsic or lightly doped body, in the DG MOSFET)

15. Due to the single operating frequency, SPDT type of switch has a limited data transfer rate. Therefore, a Double-Pole Double-Throw (DPDT) switch is designed to solve the problem
16. The DPDT switch has dual antenna and dual ports, one port for transmitting and the other for receiving, which is not sufficient for MIMO systems. Hence, we design a DP4T switch to enhance the switch performance for MIMO applications
17. A traditional DP4T n-MOS switch uses n-MOS as transistors in its main architecture and requires a control voltage of 5.0 V, while the proposed device has low control voltage
18. The switch is designed to be part of the microwave applications for switching system between the transmitting and receiving modes
19. The isolation of switch is improved by grounding RF signals on the side which is turned OFF
20. With scaling device dimensions and increasing short channel effects, multiple gate transistors can be investigated to obtain an improved gate control
21. Achieving high isolation in the OFF-state and low insertion loss in the ON-state for wideband applications is quite a challenge for switch designers
22. The channel resistance of a switch must be limited to less than about 6 Ω to achieve a low frequency insertion loss of less than 0.5 dB on a line with 50 Ω matched impedances at the source and load
23. More than 50 % of footprint saving are achieved as compared to two single-gate MOFETs, less pick and place effort
24. Energy stored is greater than 1.4 times for CSDG MOSFET compared to CSSG MOSFET
25. Replacement of traditional SiO_2 gate dielectrics with HfO_2 is mainly for requirement of high-k dielectrics
26. To improve the integration and performance of CMOS devices, and its applications, Hf-based gate layers are being integrated into DG MOSFET to achieve low leakage current
27. Excellent gate transistors with improved performance based on Hf-based gate dielectrics as the insulating layers are achieved
28. For the purpose of RF switch, we have achieved the process to minimize control voltage and minimization of the resistance for the switch-ON condition
29. Since the HfO_2 has melting point 2,812 °C, the designed DP4T switch can work sufficiently for high-power switches, Jacket water temperature, and process temperature and also for broadband and carbon nanotube-based nonvolatile RAM
30. Image acquisitions are used to observe the edges of the rectangular and cylindrical devices
31. The proposed model emphasized on the basics of the single image sensor for two-dimensional images of three-dimensional devices
32. Using image acquisitions technique, we can verify the basics of the circuit elements parameter required for the radio frequency subsystems of the integrated circuits

About the Authors

Dr. Viranjay M. Srivastava received the Doctorate (2011) in the field of RF Microelectronics and VLSI Design from Jaypee University of Information Technology, Solan, Himachal Pradesh, India, the Master degree (2008) in VLSI design from Center for Development of Advanced Computing (C-DAC), Noida, India, and the Bachelor degree (2002) in Electronics and Instrumentation Engineering from the Rohilkhand University, Bareilly, India. He was with the Semiconductor Process and Wafer Fabrication Center of BEL Laboratories, Bangalore, India, where he worked on characterization of MOS devices, fabrication of devices, and development of circuit design. Currently, he is an Assistant Professor in the Department of Electronics and Communication Engineering at Jaypee University of Information Technology, Solan, Himachal Pradesh, India. His research and teaching interests include VLSI design and CAD with particular emphasis on low-power design, Chip designing, VLSI testing and verification.

He has more than 10 years of teaching and research experience in the area of VLSI design, RFIC design, and Analog IC design. He has supervised various B. Tech. and M. Tech. theses. He is a member of IEEE, ACEEE, and IACSIT. He has worked as a reviewer for several conferences and Journals both national and international. He is author of more than 50 scientific contributions including articles in international refereed Journals and Conferences and also author of two Books, (1) VLSI Technology and (2) Characterization of C-V Curves and Analysis, Using VEE Pro Software: After Fabrication of MOS Device.

Prof. Ghanshyam Singh received Doctorate (2000) in the field of Electronics Engineering from Indian Institute of Technology, Banaras Hindu University, Varanasi, India. He was associated with CEERI, Pilani, and Institute for Plasma Research, Gandhinagar, India, where he was Research Scientist. He also worked as Assistant Professor at Nirma University of Science and Technolgy, Ahmedabad. He was Visiting Researcher at Seoul National University, Seoul, Korea. At present, he is the Professor in the Department of Electronics and Communication Engineering, Jaypee University of Information Technology, Solan, India. His research and teaching interests include RF/Microwave Engineering, THz radiation and its applications, next generation communication systems (OFDM and Cognitive radio), and Surface Plasmons and nanophotonics.

He has more than 13 years of teaching and research experience in the area of Microwave, THz radiation, Communication and nanophotonics. He has supervised various Ph. D. and M. Tech. theses. He has worked as a reviewer for several conferences and Journals both national and international. He is author of more than 170 scientific contributions including articles in international refereed Journals and Conferences.

Index

A
Absorption, 38, 145, 146, 179
Advanced modeling, 6, 9, 11, 111, 166, 172
Alternating current (AC), 9, 24, 28, 29, 94
Analysis, 6–8, 11, 36, 46, 47, 53, 73, 106, 112–117, 120, 130, 145, 156–157, 160, 169–172, 179–181
Appearance, 144, 170
Aspect ratio, 15, 47, 61, 63–66, 75, 96–97, 99, 106, 128, 143, 150
Avalanche breakdown, 8, 13, 25

B
Band gap energy, 16, 143, 144
Band gap of silicon, 16, 25, 143, 159, 181
Bands, 2, 3, 6, 9, 12, 16, 23, 26, 30, 32, 51–53, 86, 94, 95, 103–104, 148, 159, 170, 181
Barrier, 9, 11, 12, 30, 46, 49, 118, 145, 160, 180
Behavior, 7, 8, 15, 47, 52, 63, 73, 96, 118, 131–133, 177
Bipolar transistor, 3, 8, 36
BJT, 29
Body bias effect, 144–145, 150
Body effect, 96, 114
Boiling point, 144
Boltzmann constant, 88, 119, 159

C
Capacitance, 5–7, 9, 11–16, 24, 25, 27, 28, 34, 47, 48, 50, 52–54, 56, 60, 61, 63, 66, 71, 73, 75, 88, 90, 94, 97, 101–106, 111–115, 117, 121, 126–129, 133, 134, 137–138, 143, 144, 147–153, 157, 159, 160, 173, 177–179
Capacitors, 5, 7, 11, 13, 16, 25, 90, 137, 143, 148, 173, 177, 179
Carrier density, 54, 121, 133, 156
Channel, 3, 25, 45, 85, 111, 143, 165, 176
  length, 13–14, 25, 36, 48, 53, 54, 62, 63, 71, 72, 90, 96, 106, 112, 114, 121, 123, 135, 137, 145, 151, 155, 156, 180, 182
  length modulation, 9
  width, 63, 72, 90, 106, 114, 121, 156
Characterization, 7, 9–13, 31, 32, 36–37, 46, 48, 50, 57, 62–70, 72, 73, 89–94, 99, 106, 112, 116–122, 131, 133, 135, 138, 145–146, 151–153, 155–160, 170, 179
Charge, 48, 50–54, 71, 72, 112, 117–121, 129–133, 138, 155, 156, 159, 160, 178, 179
Charge model, 15, 47, 119, 136
Complementary metal oxide semiconductor (CMOS), 1, 23, 45, 88, 111, 143, 166, 175
  inverter, 6, 11, 29–31, 33–34, 36
  switch, 24, 32, 61, 63, 106
  technology, 1, 3–5, 7, 10, 11, 23, 26, 32, 34, 51, 53, 115
Complementary MOSFET, 15, 31, 53
Components, 4, 6–8, 13, 26, 31, 38, 51, 85, 111, 122, 131, 169, 170
Conductance, 62, 135
Conductivity, 25, 133
Configuration of switches, 31–34
Contact resistance, 12, 34, 75, 94, 178
Conveyor belt, 167
Cross talk, 15, 28, 111, 132–134, 138, 179
Current, 4, 23, 45, 85, 111, 143, 166, 175

C-V characteristics, 54, 121
Cylindrical surrounding double-gate (CSDG) MOSFET, 6, 9, 15–17, 111–138, 165–167, 172, 178, 179
Cylindrical surrounding single-gate (CSSG) MOSFET, 111, 121, 123, 127–129, 135–137, 178

D
1-dB Compression point, 38, 86, 87, 128
DC. *See* Direct current (DC)
Debye length, 144, 159–161
Density, 10, 14, 45, 47, 52–54, 61, 63, 71–73, 91, 99, 111, 112, 120, 121, 129–133, 135, 138, 144, 146, 156, 159, 177, 179
Depth, 38, 113, 118, 123
Device, 1, 23, 45, 86, 111, 143, 166, 175
Device structure, 45, 47, 51, 62, 66, 146, 172, 178
Device under test (DUT), 13, 72, 156, 170
DFT. *See* Discrete fourier transform (DFT)
DG MOSFET. *See* Double-gate MOSFET (DG MOSFET)
DIBL. *See* Drain induced barrier lowering (DIBL)
Diffusion, 49, 54, 71, 90, 113, 114, 121, 156
2-Dimension (2-D), 15, 47, 52, 61, 112, 113, 115, 118, 119, 122, 135, 179
3-Dimension (3-D), 11, 179
Direct current (DC), 7, 24, 28, 32, 62, 90, 132, 135, 159
Discrete fourier transform (DFT), 169
Distortion, 5, 6, 8, 10, 23, 29, 34, 36, 40, 47, 88, 90, 154, 170
Distributed, 53, 147
Double-gate MOSFET (DG MOSFET), 6, 7, 9, 13–17, 29, 33, 35, 45–76, 85–107, 111, 112, 115–121, 128, 135, 145–148, 153–160, 165, 166, 171, 172, 177–179, 179
DPDT switch, 6, 7, 32, 33, 36, 91, 106, 154
DP4T switch, 6, 7, 33, 34, 36, 37, 39–40, 61, 86, 90, 153–160, 179
Drain, 4, 25, 46, 89, 111, 143, 175
  current, 6, 15, 31, 37, 47, 52, 54, 57, 61, 62, 64, 65, 67, 69, 71, 72, 75, 76, 89–92, 94, 106, 111, 114–117, 119, 121, 124, 127, 128, 130, 132, 135, 138, 143, 145, 149–150, 153, 156, 161, 177, 178
  to source current, 61, 62, 71, 122, 153

Drain induced barrier lowering (DIBL), 9, 46, 49, 62, 71, 113, 114, 118, 119, 135, 145, 155
Drain-source, 4, 8, 16, 25, 46, 53, 66, 96, 113, 115, 150, 180
DUT. *See* Device under test (DUT)

E
Effective gate oxide capacitance, 13
Effective voltage, 11, 111
EHF. *See* Extremely high frequency (EHF)
EKV model, 9
Electric field, 8, 9, 25, 49–51, 71, 72, 113, 115, 155, 156, 180
Electron, 25, 71, 72, 119, 129, 148, 155, 156, 180
Electron charge, 119, 132, 133
Electron-hole, 30, 180
Equivalent circuits, 7, 50, 178
Extremely high frequency (EHF), 2
Extremely low frequency (ELF), 2
Extreme ultraviolet (EUV), 2

F
Fabrication, 4, 6, 9–11, 14, 16, 24, 46, 48, 61, 63, 117–118, 129, 135, 143, 147, 166, 179
Fall time, 54, 71, 90, 91, 103, 121, 127, 145, 151
Fermi level, 96
Fermi potential, 119, 129
Field effect transistor (FET), 4, 14, 135
Figure of merit (FOM), 13, 144–145, 151
Filter function, 169, 170
Flat-band, 106, 145, 159, 178, 179
Flicker noise models, 131
FOM. *See* Figure of merit (FOM)
Fully-depleted, 52, 113, 118, 145

G
GaAs FET switch, 24
Gain cascade, 86
Gamma rays, 2
Gate, 7, 25, 45, 88, 111, 143, 166, 175
  area, 14, 94, 130, 159, 166
  capacitance, 16, 63, 66, 71, 75, 90, 144
  to source voltage, 28, 36–37, 64, 65, 121, 149
  voltage overdrive, 8, 9, 54, 61, 71, 75, 89, 90, 117, 121, 151, 159, 161, 178

Index

Gate-channel, 62, 135
Gate-drain, 4, 46, 51, 131
Gate-source, 4, 25, 46
Gate-substrate, 31
Gigahertz (GHz), 1–3, 6–12, 24, 25, 32, 34, 73, 75, 85, 86, 88, 90, 91, 94, 97, 99, 100, 103–105, 125, 145, 157, 178

H

Hafnia, 16, 143
Hafnium dioxide (HfO$_2$), 9, 16, 17, 143–161, 179
Hard X-rays, 2
High dielectric constant (High-k), 16, 143–145, 147–149, 160, 180
High-field effects, 24, 89
High frequency (HF), 4, 23–25, 28, 37, 39, 50, 101, 131, 132, 144, 157, 169, 180
Hot carrier effects, 62, 112, 113, 135, 180

I

ICs. *See* Integrated circuits (ICs)
IF. *See* Intermediate frequency (IF)
Image
    analysis, 171–172
    enhancement, 170–171
    sensor, 166–169, 172, 173, 179, 181
Induced charge, 8
Inductors, 24, 26
Insertion loss, 6, 7, 10, 13, 23–25, 28, 32, 33, 38, 40, 56, 61, 72, 73, 94–97, 101, 103, 105, 106, 123, 128, 145, 156–158, 160, 177
Integrated circuits (ICs), 4, 9–12, 14, 15, 24, 40, 45, 63, 90, 111, 137–138, 145, 160, 173, 177, 178
Intermediate frequency (IF), 3, 26
Intrinsic capacitance, 6–7, 52
Inverse discrete fourier transform, 170
Isolation, 6, 7, 10, 13, 23–25, 28, 32, 34, 36, 38, 41, 56, 58, 60, 63, 77, 85, 88, 90, 94, 97, 101–103, 106, 125, 127, 128, 135, 138, 154, 177–179

J

Junction, 9, 11, 24, 25, 30, 39, 45, 51, 53, 56, 94, 101, 111, 114, 115, 123, 128, 132, 160
Junction capacitors, 25

L

Large-scale integration (LSI), 1
Layout, 7, 10, 14, 15, 28, 37, 53, 54, 56, 59, 62, 66, 67, 73, 90, 91, 113, 121, 177
Leakage current, 16, 31, 41, 47, 49–53, 56, 66, 114, 115, 119, 131–132, 143, 144, 146–148, 177
Linearity, 4, 5, 7, 13, 24–26, 38, 39, 87, 97
Long channel, 15, 47, 89, 114, 117
Low frequency, 2, 28, 97, 153, 161, 167
Low noise amplifier, 26, 36, 155
LSI. *See* Large-scale integration (LSI)

M

Medium frequency, 2
Melting point, 144, 147
MEMS, 25–26
MEMS switch, 25–26
MESFET, 4, 24–25
MESFET switch, 24–25
Metal interconnect, 189
Metallization, 118
Metal-oxide-semiconductor FET (MOSFET), 4, 24, 45, 85, 111, 143, 165, 175
Micrometer (μm), 2, 10–12, 18, 32, 54, 61, 63, 71, 89, 90, 94, 96, 98–100, 106, 117, 121, 129, 145, 146, 177, 179
Mid infrared, 2
Mobility, 6, 9, 25, 31, 46–48, 51, 54, 72, 89, 96, 112, 113, 115, 117–119, 121, 128, 133, 144, 149–151, 153, 156, 159, 160, 166, 178–180
Model, 4, 34, 46, 94, 111, 146, 166, 175
Modeling, 7–13, 46, 52, 111, 115, 135, 138, 159, 177, 179
Molar mass, 144
Molecular formula, 144
MOS capacitor, 11
MOSFET switch, 4, 6, 25, 27, 62, 73, 96, 157, 178
MOS model, 9

N

Nanometer, 10, 62, 145, 160
Near-infrared (NIR), 2
Near ultraviolet, 2
NIR. *See* Near-infrared (NIR)
Noise
    Figure, 88
    model, 131–132
    sources, 131
n-type MOS transistor, 29–30

O

Off-isolation, 13, 101, 106, 177
3rd Order Intercept Point (OIP$_3$), 39, 87
Overlap capacitances, 114
Oxidation process, 7–8
Oxide, 8, 11, 13, 15, 30, 35, 48–52, 54, 61, 63, 89, 111–114, 117–119, 121, 129, 131, 132, 136, 138, 143–145, 147–150, 152, 153, 155, 159, 173, 177–180
Oxide capacitance, 121, 144–145, 149, 150, 153, 159

P

Parameter, 9, 50, 89, 90, 95–97, 114, 116–117, 131, 137, 151, 160, 166, 171–173, 177–179
Parameter extraction, 9, 50
Partially-depleted (PD), 33
Path loss, 85–86
Phase noise, 88, 131
PIN diode switch, 24
Pixels, 165, 166–167, 169–171
Poisson's equation, 49, 112, 113, 118, 119, 121, 129
Polysilicon, 9, 138, 173, 177, 178
Post processing, 168
Potential, 4, 5, 9, 11, 14, 15, 31, 32, 41, 47–49, 71, 73, 111–115, 117–120, 129, 132, 138, 155, 160, 173, 177, 179, 181
Power
  amplifier, 3, 8, 26, 27, 35–36, 88, 145, 155
  dissipation, 8, 26, 29, 36, 39, 50, 63, 73, 145, 153, 178
Pre processing, 167–170
Propagation delay, 39, 40, 101, 145, 152, 152, 177
p-type, 4, 13, 24, 27, 28, 30, 31, 36, 37, 53, 56, 59, 88, 96, 101, 102, 114, 115, 117, 146, 149, 157, 159, 180
Punch-through, 46

R

Radio frequency (RF), 1, 3, 5–7, 9–13, 15, 23, 26, 29, 36, 61, 85, 111, 137, 143, 145, 160, 173, 177, 178
  MOSFETs, 7–8
  power handling, 38–39
  switches, 3–5, 7, 9, 10, 15, 25, 31, 32, 75, 90, 97, 145, 178
  switch performance parameters, 37–39
  transceiver switch, 28–29
  transceiver systems, 26–28

Recombination, 30
Regime, 7, 10, 11, 15, 24, 25, 45, 47, 52, 62, 118, 178, 179
Resistances, 6, 15, 25, 50, 53, 60, 90, 101, 111, 121, 122, 128, 131, 138, 143, 150, 160, 173, 177
Resistivity, 11, 24, 111, 122, 132, 133, 160
Resistors, 28, 94, 96, 131
Return loss, 13, 38, 90, 94, 95, 104, 105
RF. *See* Radio frequency (RF)
Rise time, 54, 71, 90, 103, 121, 127, 145, 152

S

Saturation, 8, 9, 15, 36, 47–49, 61, 71–72, 113, 115, 119, 138, 148, 154–156, 179
Scattering, 49, 51, 89, 133
Self-alignment, 52, 112
Semiconductor, 8, 10, 15, 24, 35, 63, 72, 96, 119, 133, 134, 143, 148, 156, 160
Series resistances, 99, 114, 158
Short channel effects (SCE), 10, 14, 15, 45–49, 53, 62, 75, 85, 99, 105, 106, 111–113, 131, 136, 145, 178
Shot noise model, 131–132
Silicon dioxide, 145
Silicon-on-insulator (SOI), 111, 145, 166
Single-gate MOSFET, 34–36, 61, 96, 97, 112, 136, 146
Single-pole double-throw (SPDT) switch, 32
Skin depth, 38
Soft X-rays, 2
SOI. *See* Silicon-on-insulator (SOI)
Solubility in water, 144
Source, 4, 6, 8, 12–14, 16, 25, 28–31, 36–37, 46, 49, 53, 54, 58, 60–62, 64–66, 71, 87, 90, 96, 97, 100–103, 111, 113–115, 118, 119, 121, 122, 124, 127, 128, 130–132, 136, 143, 149, 150, 153–155, 180
S-parameters, 103–106, 178
SPICE, 124, 125
SPST switch, 31
Strong inversion, 48, 53, 118–120
Substrate, 6–9, 11, 13, 14, 16, 17, 25, 26, 28, 30–32, 48, 53, 56, 58, 89, 94, 96, 97, 101, 102, 111, 112, 114, 116–119, 123, 125, 131–134, 136, 147–149, 152, 159, 160, 165, 180
Substrate current, 6, 8, 11, 13, 14, 30, 31, 48, 61, 111, 129–130
Subthreshold, 15, 45, 47, 49–51, 112–114, 117, 118, 136, 146, 179, 180
Subthreshold swing, 49, 113, 118, 180

Index

Super high frequency, 2
Super low frequency, 2
Switching speed, 6, 8, 13, 15, 24, 26, 39, 40, 90, 91, 94, 103, 106, 111, 143, 149, 160, 177, 178
Switch topology, 32

T

Thermal noise, 87–88, 131, 132
Thermal noise model, 131
Third harmonic, 87
Threshold voltage, 15, 28, 36, 47–49, 51–53, 56, 63–66, 76, 90, 96, 106, 107, 111–115, 117, 118, 122, 138, 145–146, 149–151, 160, 166, 177, 178, 180
TLF. *See* Tremendously low frequency (TLF)
Transceiver systems, 1–6, 26–28, 36, 85, 86, 89
Transconductance, 8, 15, 36, 48, 61, 62, 71, 89, 98, 116, 117, 132, 135, 154, 155, 157
Transition time, 39
Transmit/receive (T/R), 4, 6, 13, 29, 159
Tremendously high frequency/far infrared, 2
Tremendously low frequency (TLF), 2

U

Ultra high frequency (UHF), 2
Ultra low frequency (ULF), 2

V

Very high frequency (VHF), 2, 3, 24
Very-large-scale integration (VLSI), 1, 11
Very low frequency (VLF), 2
Voltage controlled oscillator (VCO), 5
V_{th} roll-off, 118

W

Width, 5, 10, 12–14, 27, 31, 32, 46–48, 50, 53, 54, 62, 63, 66, 71, 72, 87, 88, 90, 94, 96–98, 106, 112, 114, 121, 127, 131, 132, 138, 155, 156
Wireless local area networks (WLAN), 2, 6, 11, 23, 101
WLAN. *See* Wireless local area networks (WLAN)
Work function, 48, 49, 51, 52, 56, 61, 62, 106, 118, 122, 130, 135, 144, 154, 180

Printed by Publishers' Graphics LLC
LMO131023.15.14.37